Alaska Gold
The History of Gold Dredge No. 8

Alaska Gold

The History of Gold Dredge No. 8

Maria Reeves

Gold Fever Press

Fairbanks, Alaska

Gold Fever Press

Reeves, Maria, 1981—
Alaska gold: the history of gold dredge no. 8

Cover Photo: Gold Dredge No. 8 as she is today.

Publisher: Gold Fever Press
Editor: Kent Sturgis
Cover and inside design: Maria Reeves
Proofreader: Melanie Wells
Printer: Sheridan Books, Inc.
Text: © 2009 by Maria Reeves

International Standard Book Number: 978-0-578-01159-2

Printed in the United States of America
10 9 8 7 6 5 4 3 2 1

Table of Contents

Dedication

This book is dedicated to my husband, Doug, and my family. Thank you all for encouraging me to write this book. Doug, thank you for your love and the sacrifices you make for our country. Mom, for being wise, noticing the small things, and pushing me to better myself. Dad, for all of your amazing stories and out-of-the box thinking; my next book will probably have to be about you! I'd also like to thank you for pointing out the difference between manganese and magnesium before going to print! Lauren for following your dreams and making things happen. Kinzey, I can always count on your hilarity... you've got my vote for governor! Jordy, thanks for putting your books down long enough to let me out of the penalty box: you've got it all. Ilaura Wuggles, 6 AM isn't as early when you're there... but it's still early! Keep living the dream, Wug. Oma, you're quite the inspiration! Nana & Bop Bop, I love you guys. To Kent Sturgis, the best guide in writing I could have asked for in this venture. To Chelsea, my favorite meepzorp. Micky, thank you for being so dependable. Beenen, college would not have been nearly as fun without you. To my swimmers on NPAC and NPHS for giving me a wonderful two years - I'll miss you all as my adventure continues. To Ms. Stitham, my favorite monster woman. And of course, the WST, always.

Introduction

An introduction from Dad:

If you're reading this in a gift shop you're probably trying to decide whether to buy this book. If you've already bought it, let me be the first to congratulate you. So, dear reader, please allow me to introduce the author of this fine read, my daughter Maria Michelle Reeves. I told Maria I would love to write the introduction if she would agree not to edit it, no matter what. She agreed. So here goes.

Imagine my delight when Maria told me she was going to write a book about Gold Dredge #8. Lots of people want to write a book but I only know a few who have done so. Maria's book is a fun and informative historical record of mining in the Fairbanks district and the massive machines that turned the earth upside down to recover the gold found buried in the earth's crust.

Maria's three sisters and brother grew up in a bunkhouse at Gold Dredge #8 where our family lived until we sold the operation to Westours in 1996. When my wife Ramona and I talked about how we were going to break the news that we were selling, we figured two of the five kids would be sad, two would be glad, and one (the youngest) would think all the brouhaha was about her. We were right all around. Maria was in the first group, as we predicted. No dad likes making his kids cry, but business is business.

However, unbeknownst to them at the time, I had arranged to purchase another dredge (#2 on Fairbanks Creek) so they would always have a place to go and pan gold and experience the awe and majesty of these engineering marvels from another era. My wife and I figured by buying another dredge, two of the five would be glad, two of them would be sad, and the little one would think all the brouhaha was about her. Yep, right again. Everything seems to balance out on the creeks. Dads like making their kids happy because it's not always about business.

We have continued our relationship with the history of the FE Company by buying the company's Fairbanks holdings, making us the largest private landowners in Alaska, behind the state and federal governments and Native corporations.

It's nice to see that Maria's interest in the history of Gold Dredge #8. Westours' commitment to make this book available to its customers eventually will contribute to an evolving fact--someday Gold Dredge #8 will have made more money for the owners in tourism than it ever did while it mined for gold. It probably already has.

Something else about the author you may find interesting: Maria is the fastest girl swimmer Alaska ever produced and still holds the state high school record in the 50-yard free style although she graduated in 2000. She competed in the Olympic trials in 2000 and attended Northwestern University on a full scholarship for swimming, graduating with degrees in Communications and English in 2004.

With all that book learning and education, she learned how to write worth a (expletive deleted) but she still hasn't learned how to swear worth a (expletive deleted). How in the (expletive deleted) did that happen? Enjoy.

--John Reeves, Maria's Dad

Chapter One

Looking Back

I wanted the gold and I got it;
I scrabbled and mucked like a slave.
Was it famine or scurvy – I fought it;
I hurled my youth into a grave.
I wanted the gold, and I got it –
Came out with a fortune last fall,
Yet somehow life's not what I thought it,
And somehow the gold isn't all.
– Robert Service

Gold Dredge No. 8 in the Goldstream Valley, just north of Fairbanks, is one of the most important reminders of the rich and colorful mining history of Interior Alaska. A giant gold-mining machine, it was one of eight dredges owned by the United States Smelting Refining and Mining Company and operated by its subsidiary, Fairbanks Exploration Company.

Dredge No. 8 operated between 1928 and 1959, playing an essential role in the extraction of gold from the Fairbanks District. Despite its name, No. 8 actually was the *third* dredge built for USSR&M. None of the dredges owned by the company were numbered one or nine.

Gold dredges are measured by the size of their buckets. No. 8 is a six-foot dredge, which means that each of its 64 buckets is capable of holding six cubic feet of dirt with every pass. There are 27 cubic feet in one cubic yard. Placer mines gererally descrive their mining operation by relating how many yards of gravel they process hourly or daily.

According to company records comparing the recovery of gold to the prospect value of the land, the dredges were able to recover 97% of the gold in the land – astonishing efficiency even by today's standards. However, the dredge was not designed to recover larger nuggets. In fact, all nuggets with a diameter larger than $1^{5}/_{8}$" washed out of the back of the dredge alongside the tailings. Today, every dredge tailing pile contains gold nuggets, though the cost of finding them usually outweighs the rewards. Tailings are often used to build Alaskan roads, which means that Alaskan roads are literally paved in gold.

After it was shut down in 1959, Gold Dredge No. 8 sat vacant more than a decade before it was purchased by a small group of Fairbanks businessmen who offered tours there. In 1973, the tours were discontinued. Once again the dredge was deserted.

In 1982, entrepreneur John Reeves purchased the dredge. Together with his wife, Ramona, Reeves restored No. 8 and improved the surrounding property by moving a bunkhouse to its current location and building a 4,000-square-foot deck overlooking the dredge. The Reeves family opened the dredge to the public in 1983. Reeves campaigned successfully to add Gold Dredge No. 8 to the National Historic Register of Historic Places. By 1984, the area had also won recognition as a historical district and the dredge became a National Engineering Landmark.

Five children and more than a decade later, Reeves had made significant improvements to Gold Dredge No. 8 and the property, having moved all of the historic buildings to their current locations. Holland America Westours bought the dredge from Reeves in 1996. Each summer, it draws thousands of visitors.

It has been more than eighty years since Gold Dredge No. 8 began operating in the Goldstream Valley. Even though she sleeps silently in a self-made pond and never was operated again, No. 8 still symbolizes the heart of the Interior's mining history.

There's gold, and it's haunting and haunting;
it's luring me on as of old,
yet it isn't the gold that I'm wanting
so much as just finding the gold.
— Robert Service

Maria Reeves

Gold Dredge No. 8 rests in a self-made pond on the outskirts of Fox, Alaska. It has been inactive since 1959 and is now owned by Holland America Westours. It is open to the public during Alaska's summer months. Thousands of visitors come to Gold Dredge No. 8 each summer to experience a genuine piece of Alaska's mining history.

Chapter Two

Stripping & Placer Mining

Gold is where you find it.
-Mining Proverb

There are two types of gold. Lode gold is found inside hard rock formations and has to be extracted by crushing the rock surrounding it. Placer deposits result from natural weathering. They can be deposited by rivers or slowly carried by glaciers. Either way, water is their main mode of transportation.[1]

Rivers carry sediment downstream, which causes them to leave behind alluvial deposits. Alluvial deposits are made of clay, quartz, miscellaneous rock, and gold.[2] When placer deposits are found above current or former riverbeds, they are called bench deposits. It is possible for placers to exist anywhere from the earth's surface to hundreds of feet below it.[3]

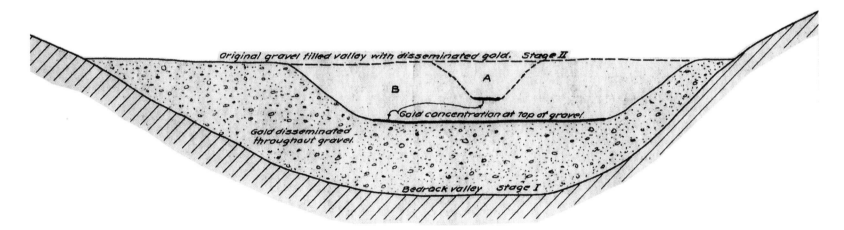

Figure 2-1 is from an unpublished report titled, "*The Fairbanks Placer Gold Deposits (With Map Folio)*" by J.M. Metcalfe and Ralph Tuck, published courtesy of the Fairbanks Gold Archive Collection. It is a cross-section illustrating the formation of a top-of-gravel gold concentration. Bedrock valley of stage 1 has, through stream aggradation, become filled with gravel from a gold-bearing source (Stage II); fine gold is disseminated uniformly through the alluvium. Lowering of base level has resulted in the carving of valley A. Valley A has a small top-of-gravel gold concentration on its valley floor as a result of the reconcentration of gold originally disseminated through the gravel. Continued lateral and down-cutting, after the formation of valley A, has carved out the larger valley B, which has a higher grade and more extensive top-of-gravel gold concentration. If, at this time, the base level was raised, valley B would be filled with alluvium and a buried gold-bearing horizon would result.

Historically, placer mining played a major role in Alaska's economy because it is one of the least expensive and most simple ways to mine. Placer mining simply involved washing gravel over sluice boxes. Even panning for gold, as Alaska's earliest prospectors did, is a form of placer mining.

Placer mines range from recreational to large-scale mines. Gold Dredge No. 8 was a large-scale operation.

Though Gold Dredge No. 8 is itself a placer mine, it never would have been successful without stripping. Stripping involves spraying a lot of water against the ground at a very high pressure. Hydraulic Giants, also known as Iron Giants, were massive water canons that the F.E. Company operated in order to strip the land ahead of Gold Dredge No. 8. During stripping, the water breaks up and slowly thaws the muck and permafrost before washing away all of the ground's overburden to expose the gold-bearing gravel. The gold-bearing gravel lies on top of bedrock. The most common type of bedrock is schist, which is sedimentary in origin. Next, the gold-bearing gravel and top few feet of bedrock are mined for gold.

OPERATION OF UNITED STATES SMELTING REFINING AND MINING COMPANY

Figure 2-1 is from a USSR&M company pamphlet published courtesty of the Fairbanks Gold Archive Collection. It depicts the two different types of mining processes employed by Dredge No. 8. First, up to 200 feet of overburden had to be removed in a process called stripping. Dredge No. 8 was a type of placer mine. Before the land was stripped, USSR&M used Keystone drills to determine the gold content of the land. Next, a route for Gold Dredge No. 8 was created based on drilling results and the stripping process began. Depending on the depth of the overburden, it could take between one and three seasons to expose the gold-bearing gravel, which then was thawed with thaw points. Once the ground was thawed, it would not freeze again. USSR&M thawed the land a year before it was dredged. They did this as a safeguard against any shortage of water, which was a crucial ingredient in the process.

Chapter Three

Fairbanks Gold

Alaska was purchased from Russia in 1867 for $7.2 million in a transaction called the Alaska Purchase. At 586,412 square miles, Alaska cost just under two cents an acre. Secretary of State William Seward engineered the purchase, which was derided by newspapers as "Seward's Folly" because it was widely unpopular. In 1959, Alaska became the forty-ninth state.

Gold is a precious metal that has been used as a measure of wealth throughout the years. Mineral production records have been kept in Alaska since 1880. By 1939, the territory had produced more than $777 million dollars in mineral wealth, more than one-hundred times the purchase price. About 66% of total came from gold, though other minerals such as copper, silver and lead also were recovered. [1]

1939 Mineral Production Chart

Gold	$512,657,000
Copper	$227,375,000
Silver	$13,484,000
Coal	$11,624,000
Lead	$2,482,000
Tin	$1,546,000
Platinum & Others	$8,645,000
Total	**$777,818, 000**

Figure 3-1 is based on information from the unpublished report, *"The Fairbanks Placer Gold Deposits (With Map Folio)"* by J.M. Metcalfe and Ralph Tuck, published courtesy of the Fairbanks Gold Archive Collection. It shows the value of gold and other minerals that had been taken out of Alaska in 1939. At the time, gold traded at $20.67 per ounce.

Placer gold accounted for 64.9% of the gold taken out of Alaska by 1939, while lode gold accounted for the other 35.1%. Combined, Fairbanks and Nome had the most productive placer deposits. Together they accounted for 68% of all placer gold taken out of the territory of Alaska. By 1939, Fairbanks alone had produced $334 million in gold at $20 per ounce, accounting for 34% of the total for the territory.[2]

During the Ice Age, the Fairbanks gravel deposits were exposed to low temperatures, and both the gravel and bedrock were frozen to a depth of 200 feet below the earth's surface.[3] Gold-bearing veins in Fairbanks occur in three main areas that center around Ester, Pedro Dome, and Gilmore Dome. Placer gold exists long after bedrock has been eroded, and is not easily scattered by natural forces due to its density. The gold in Fairbanks placer deposits dates back to the Tertiary Period and is still very close to the original source.[4]

Placer gold results from a bedrock source such as quartz veins. Most of the gold delivered to a stream is freed from bedrock by weathering, not the stream's corrosion. Typically, these weathering processes are caused by gravity, which manifests itself in processes called "sheet wash," creeping, or sliding. Once the gold appears in the stream bed, its transportation is contingent on the grade of the stream, the volume of water running through the stream, the amount of debris to be found in the stream, and the size of the gold. Flake gold, which is very fine, can be transported many miles. Coarse gold, which is defined as pieces of gold weighing more than one milligram, is never found far from its original source of bedrock. In fact, coarse gold is rarely found greater than three miles from its source. In order for coarse gold to travel a distance of 10,000 feet, a stream must slope between 2,000-4,000 feet.[5]

Gold discovered in the Goldstream Valley has two sources. Pedro Dome feeds it from tributaries on the right limit, while Gilmore Dome feeds it from tributaries on its left limit. The gold recovered from the Goldstream Valley has the same chemical makeup as these two sources.[6] The tributaries in Goldstream contribute to gold distribution in the area and are responsible for any abnormal jumps in value.[7] Most of the gold in the Fairbanks district occurs as a bedrock concentration in the lower five to ten feet of gravel. If the bedrock has been weathered, it can cut several feet into the bedrock.[8]

More than 80% of the world's gold is used to produce jewelry.[10] The purity of gold is measured in karats, with twenty-four karats designating pure gold. Lower karat numbers designate less pure gold. Chemically pure gold does not occur naturally, though it can be created. Pure gold is both the most ductile and malleable of all metals. Pure gold can be beaten to a width of .00001 mm[11] and gold leaf can be beaten until it becomes translucent.[12]

While pure gold is malleable and easy for a jeweler to work with, most jewelry stores in America do not sell twenty-four karat gold because what is easily formed is easily deformed. This is why jewelers combine gold with other metals to create stronger alloys. Even though fourteen karat gold is less pure than twenty-four karat gold, fourteen karat gold is much stronger and more suitable for jewelry. By and large, most American jewelers sell ten, fourteen and eighteen karat gold jewelry.[13]

Gold alloys vary in color depending upon the type of metal with which it has been combined. Copper and gold alloys create a pink color. Silver and gold alloys create a green color. Nickel and gold create a white color. Silver, copper, and gold alloys create a traditional yellow color.[14]

Q: What weighs more, a pound of gold or a pound of feathers?

A: While most people would guess that they weigh the same, the answer is that a pound of feathers weighs more than a pound of gold.

The reason? Gold is measured in troy ounces, instead of avoirdupois ounces. A troy ounce is a unit of measurement that is only used to describe the mass of precious metals such as gold, silver, and platinum. There are twelve troy ounces in one troy pound while there are sixteen ounces in an avoirdupois pound.

It is important to note that placer gold is never pure. Occasionally placer gold can contain as much as 40% impurities. In Fairbanks, gold purity ranged from 800 to 950 parts per thousand. Mechanical impurities are usually the type of rock that the gold occurred in before it was freed from its source. Most of the impurities found in Fairbanks gold were quartz, because the gold most often originated in quartz veins. Coarse gold can have more quartz in it because it doesn't move through streams as quickly as smaller flakes, and it takes longer for them to erode. When quartz or bedrock is visible on the gold, it is good evidence that the source is nearby. A lack of impurities is evidence to the contrary. Silver is the most common type of chemical impurity found in placer gold in the Fairbanks district and it usually occurs between 50 to 200 parts per thousand.[9]

Maria Reeves

This bracelet was handcrafted at *Gold Rush Fine Jewelry* in Fairbanks, Alaska. It is a beautiful example of three different gold alloys at work in a single piece of jewelry. The rose gold detail is a gold and copper alloy. The white gold detail is a gold and palladium alloy. The yellow gold band is a gold, silver, and copper alloy. The natural gold nugget in the center was recovered in the Brooks Range, north of Fairbanks. Visitors are welcome to stop by the workshop at *Gold Rush Fine Jewelry* and watch firsthand as professional goldsmiths create unique jewelry designs using natural Alaskan gold nuggets.

Chapter Four

Fairbanks Booms & Busts

A gold miner is a liar standing beside a hole in the ground.
-Samuel Clemens

After the 1896 strike in Canada attracted thousands of stampeders north to the Klondike, thirty-four separate gold stampedes took place in Alaska.[1] The city of Fairbanks has seen its fair share of booms and busts tied to gold. In fact, Fairbanks already had been through several boom-and-bust cycles before the dredges began operating.

In the summer of 1901, an Italian immigrant named Felix Pedro and his partner, Tom Gilmore, stopped to rest on a hill that is now named Pedro Dome. The pair had started a 165-mile trek to Circle City to purchase supplies. While resting, Pedro, whose Italian name was Felice Pedroni, noticed smoke rising from the Tanana Valley below. The smoke was coming from a sternwheel riverboat trying to navigate the Bates Rapids on the Tanana River about twenty miles away. Pedro and Gilmore began making their way toward the riverboat hoping to purchase supplies and save them the long walk to Circle City.[2]

Little did Pedro know that while he and Gilmore moved to intercept the struggling sternwheeler, Captain E.T. Barnette was desperately trying to get upriver to the Tanana Crossing, where gold-seekers following the Eagle Trail from Valdez to Eagle forded the Tanana River. Barnette and his wife, Isabelle, traveled with $20,000 worth of trade goods, hoping to set up a trading post at the crossing. Their Captain was a man named Charles Adams and the boat they were on was the *Lavelle Young*. Unfortunately for the Barnettes, low water in the Tanana River had made the rapids impassable, forcing the *Lavelle Young* to backtrack into a nearby tributary, the Chena River. On August 26, 1901, after a heated debate with Adams about whether it would be possible to continue upriver, Barnette reluctantly set up a temporary trading post on the banks of the Chena River near what is now the intersection of First Avenue and Cushman Street in downtown Fairbanks.[3]

It took about a day for Pedro and Gilmore to reach the *Lavelle Young*. When they arrived, they were pleased to find Barnette and his crew busy unloading freight. Pedro and Gilmore brought good news with them: they were not the only prospectors in the area in need of supplies.[4] Pedro and Gilmore stocked up on bacon, flour, and beans before resuming their search for gold.[5] Things were finally turning around for Barnette. Soon, they would turn around for Pedro as well.

On July 22, 1902, about a year after his chance encounter with Barnette, Pedro discovered gold on Pedro Creek about sixteen miles northeast of Fairbanks. Pedro's discovery became the tenth significant strike in Alaska. It also caused several stampedes, which in turn led to even more discoveries. Soon camps were established on Goldstream, Cleary, Ester, Dome, Eldorado, and Fairbanks creeks.

Fairbanks was incorporated in 1903. Federal District Judge James A. Wickersham, who had moved his court to Fairbanks from Eagle after the discovery of gold, persuaded Barnette to name the town in honor of Wickersham's mentor, U.S. Senator Charles Fairbanks of Indiana. In 1904, Senator Fairbanks was elected vice president under Theodore Roosevelt.[6] That same year, construction began on what became the narrow-gauge Tanana Valley Railroad that connected many of the camps.

Following Pedro's discovery, stampeders used several methods to extract gold. Often, miners sunk vertical shafts, called drift mines, into bedrock. In most cases, miners had to sink these vertical shafts during the winter. The ground was thawed with steam and lifted to the surface, where it was dumped out and stored. In the spring, when the creeks thawed, the miners sluiced through all of the material that they had stored during the winter.

Barnette decided to stick around the boom town and see what he could make of himself. But he was impatient. At one point, he salted a mine to start another stampede. He also wrote letters to anyone he could think of exaggerating how much gold was waiting to be found.[7] One such letter read, "I have just returned from No. 2 Above, on Pedro. The owners wanted $2,000 of supplies, so I went over to make sure their ground was good. I found seven holes down, but not to bedrock…They have already six feet of coarse gold that will average 15 cents a pan. I found much better than that. It was the best I ever saw…What I have said in this letter are facts, and the man that gets here early is in it."[8]

In 1902, three days after Christmas, Barnette sent his cook, Jujiro Wada, to Dawson City to stir up the locals with news of Pedro's strike. Wada embellished his account of the strike to such a degree that nearly 1,000 men left Dawson City, traveling 300 miles to reach Barnette's trading post in the dead of winter, even though the temperatures ranged anywhere from 50 to 74 degrees below zero.[9]

By 1903, activity around Fairbanks declined as miners realized that most of the prospects around Pedro Creek already had been claimed. Many stampeders found themselves trapped for the winter without enough money or food to last through spring. They were angry, blaming Barnette and Wada for their misfortunes. Wada was tried in a mock trial. Wada's explanation was that when he went to Dawson, the situation in Fairbanks had been as he described. Even Barnette acknowledged that likely there would not be a large cleanup that year. By June of 1903, many stampeders had left Fairbanks for greener pastures.[10]

That fall, however, several big strikes were made. The first was on Fairbanks Creek, where the richest single pans were known to yield an astonishing $130 worth of gold. Fairbanks Creek would become the richest strike in all of Alaska. The second was on Cleary Creek, where the original gold streak went from top to bottom of a lucky miner's drift. The third strike took place on Ester Creek, but was kept a secret for eight months by its discoverer, John Mahilcik. Being in the right place at the right time, Barnette would make millions, as would the miners with claims on Cleary, Fairbanks, and Ester creeks.[11]

UAF Archives

A photo of Felix Pedro. In 1898, Pedro and his partner discovered the richest creek they had ever seen somewhere in the seemingly endless Tanana Hills. Though they marked the creek with an upside-down boat, they were never able to find it again. In 1901, Pedro and a man named Bert Johnson found gold at a stream they named "Ninety-Eight Creek." Pedro believed it was his long-lost creek, so he and Johnson returned to Circle City to get supplies and share the news. Unfortunately, the creek proved to be worthless and Pedro was nicknamed the "Old Witch" by the other miners. But Pedro refused to give up. Soon after his second setback, Pedro partnered with Tom Gilmore and continued his search.[12]

In his book, *Crooked Past*, Terrence Cole compared Barnette and Wada to an unnamed poem by R.N. De Armond. The poem is about a gold miner named Joe Smith who cannot get through the Pearly Gates of Heaven because it is too crowded inside.

In a stroke of genius, Smith asks Saint Peter to tell an old mining buddy that there has been a placer strike in Hell. Once the miners have begun pouring out of Heaven, Saint Peter finally invites Smith inside. But, once Smith is faced with the opportunity to enter Heaven, he grabs his pack and bolts off after the other miners... just in case there really is a placer strike in Hell.

The Saint appeared and, nothing afeared,
Joe said he would like to send
to Jack McAdoo, of Caribou,
a message straight from a friend.
"Make no mistake, it's 'Jitney Jake,'
he's the only one you'll tell;"
and drawing him near, hissed in his ear,
"there's a placer strike down in Hell!"

With a glow of content, then back he went
to his seat on the Outer Rim.
They were all alike, and he knew he'd hike,
if such a message had come to him.
And soon they sped, with a stealthy tread,
some he thought he never would see;
they were of every date, from fifty-eight
to nineteen and twenty-three.

At last to the door came Peter once more,
and he said there was room to spare;
but naught heeded Joe, as he muttered low:
"Gosh! Maybe there's something there!"
So with eyes ablaze in an eager gaze,
and the look of a man possessed,
he hoisted his pack on his old bent back,
and hurried off after the rest!
–R.N. De Armond[13]

In 1903, gold production in the Fairbanks District totaled $40,000. A year later, gold production reached $600,000. By 1906, the area's mines yielded $9 million worth of gold. In 1910, more than $30 million worth of gold was produced on Cleary, Fairbanks, and Ester creeks alone.[12] Meanwhile, Alaska's largest year of gold production had occurred in 1906 when more than one-million troy ounces were mined. In 1906, one-million troy ounces of gold was valued at more than $22 million dollars.[14] In 2008, that same amount would be worth more than $1 billion!

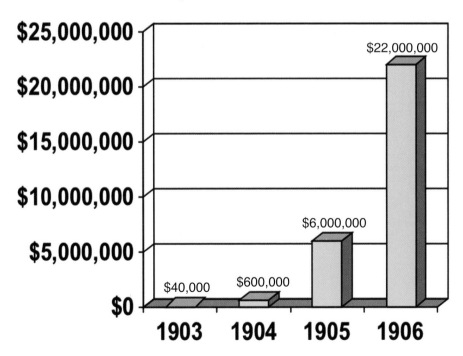

Gold Production In Fairbanks 1903-1906
At $20.67 Per Ounce

Figure 4-1 was compiled based on information published in *"Alaska Geographic: Rich Earth: Alaska's Mineral Industry."* It is a graph depicting the original value of gold produced in the Fairbanks District between 1903 & 1906.

In 1910, Fairbanks was a booming town with an official population of 3,541 people. Many miners who lived near their claims were not included in the census because they resided outside city limits. It was estimated that 11,000 people lived in or around Fairbanks. Wickersham made Fairbanks the administrative center of the Third Judicial Division. In 1917, The Agricultural College and School of Mines was created.[15] Eventually, this college became the University of Alaska.

In 1909, gold production in Fairbanks began to taper off.[16] In the years that followed, many left Fairbanks in search of better prospects. In 1917, at the onset of World War I, many more men left Fairbanks to fight in the war. It is estimated that 10,000 people left Fairbanks from 1911 to 1923.[17] By 1920, most of the gold accessible through drift mines had been exhausted, and few mines were still active. In 1923, construction of the Alaska Railroad was completed, with Fairbanks as its northern terminus, but it would take more than a railroad to revive the rapidly shrinking town.[18]

USSR&M became interested in the Fairbanks district in 1924, well after Alaska's peak year of production. At the time, it was commonly believed that the land in the Fairbanks district had been exhausted. The F.E. Company was just what Fairbanks needed. It brought the town back to life with hundreds of new jobs and economic stability that followed its investment in the construction of the Davidson Ditch.

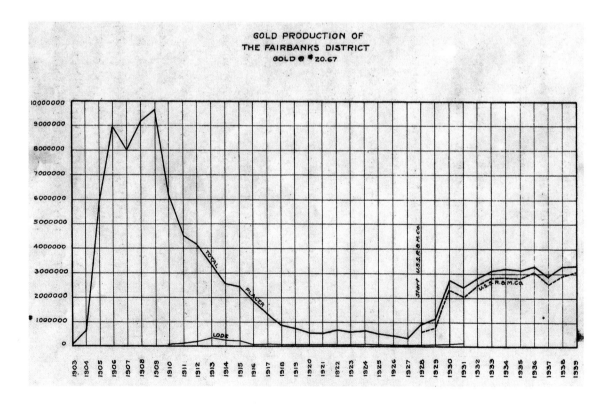

Figure 4-2 is from the unpublished report, *"The Fairbanks Placer Gold Deposits (With Map Folio)"* by J.M. Metcalfe and Ralph Tuck, published courtesy of the Fairbanks Gold Archive Collection. It is a graph of all gold production in the Fairbanks district between 1903 and 1939. Gold production steadily declined until 1928, when the Davidson Ditch was completed and the first USSR&M dredges began operating.

Chapter Five

Davidson Ditch

Placer mines require a large volume of water to separate gold from other materials. The process takes advantage of the fact that gold is nineteen times heavier than water, while regular soil is only about four times heavier. Water is used to wash the gold-bearing soil and gravel over riffles and, because gold is so dense, it gets caught in riffles while the other materials wash over them. If water is not available on gold-bearing land, it must be brought in. This can be an expensive undertaking.

In 1898, James M. Davidson traveled north to the Klondike after losing almost everything during the finacial panics of the 1890s. Once he had crossed the Chilkoot Pass, Davidson realized that he did not approve of Canadian mining methods or laws. He decided to move to Alaska and settled near Circle.[1] Davidson had a degree in civil engineering from the University of California.[2]

Upon hearing about the gold strikes in Nome, Davidson booked the first riverboat passage of the season. He arrived in Nome on the Fourth of July in 1899. Davidson used every last dollar he had to purchase a lot and a tent. He then started the first surveying transit in the city of Nome. By 1901, Davidson had noticed that water useable for mining was quite scarce. Davidson formed the Miocene Ditch Company with two other men, W.L. Leland and W.S. Bliss. The ditch was completed by 1904, at which point Davidson began visiting every single placer mining district in Alaska.[3]

In 1921, Davidson had a plan for the Fairbanks district. He envisioned a large-scale dredging operation. He wrote Norman C. Stines, a mining engineer, that he believed water could be supplied through a ditch originating in the Chatanika River. Davidson spent the next two years studying the area and purchasing options in Fairbanks. He even staked out a preliminary route for the ditch that was nearly 100 miles in length.[4]

Stines was with the Miami Corporation when Davidson wrote him. Stines responded in a letter dated November, 15 1923, "I am decidedly interested in what you have to say, and can tell you now that I am going to plan my work here so that I can meet you in Chicago about the 25th of this month… I appreciate greatly what you have said and want a chance to work with you on this. As you have written me confidentially, I will do the same with you. I am not sure that my present people will undertake such a proposition, but I am certain that I can get others who will… I am seriously considering severing my present connection and going back as a free lance as I was when I went to Nome in 1920."

Stines asked Davidson to bring all of his information to their meeting in Chicago. He also asked that Davidson be prepared to allow him to take the data with him to study when he left Chicago.

In 1923, Stines was convinced that dredging could be profitable in the Fairbanks district. Stines was backed by the USSR&M Company when he purchased Davidson's options and began prospect drilling.[5] According to company records in the Fairbanks Gold Archive Collection, Davidson sold his options to USSR&M for 5% of the net profit allowing for all expenses such as purchases, interest charges, income taxes, etc. The contract with USSR&M called for payments in the sum of $6,000 annually in quarterly installments of $1,500 until USSR&M began "to realize on his profits." Payments began on March 31, 1924. Once Stines was convinced of the project's success, he was fully committed. Charles Janin, a notable mining engineer of the time, recognized Stines as an "ambitious engineer with a tendency towards promotion and grandiose plans."[6]

USSR&M was not without competition. In 1922, Wendell Hammon of the Hammon Engineering Company hired California engineer Ray Humphrey to purchase mining options in Fairbanks. Fifteen years earlier, Hammon had proposed bringing water from the Chatanika River to Fairbanks but his proposal was rejected due to the high cost.[7]

In 1923, Hammon purchased the Pioneer Mining Company, based in Nome. Hammon's latest business venture included dredging fields at Bonanza and Eldorado creeks. Hammon sent several experienced prospecting crews to Fairbanks to determine whether dredging could

Mr. Norman C. Stine,   Nov. 7th. 1923.   File........1-8-2-18
#137 Lasalle Street
Chicago, Ill.

                    Dear Mr. Stine:

                              On my return from Alaska, at Seattle
a few days ago, I receive a letter from you, dated at Torronto,Canada,
in which you enclosed copies of letters written to John Hopp, as per
the address I gave you last spring, when I was going north that you
mig ht communicate with him directly, during my absence, concerning
his hydraulic and dredging property. I called Mr. Hopp up immediately
and asked him to come down to my hotel and tell me about it which he
did: he said he answered your letter at the time it was received, but
not hearing further from you he had taken it up with other parties
who were on the ground now making investigations. I am sorry about it
but it apparently was no ones fault; the answer probably miscarried
thus terminating the correspondence.

                    I am now personally quite anxious to see you and talk over
a mining proposition, that I think is a wonder if only enough money
can be interested to put it over in the shape it should be, but its
no poor man's proposition; it is something which will take a lifetime
to work out, in a very rich placer country, and wher the ground can
be handled very cheaply, if only sufficient capital can be interested
to properly equip the same. The values as has been determined by the
miners are very high, but it would be necessary for to thoroughly drill
the property before investing in it all. It will be a dredging property
where a number of dredges may be installed and work at the same time,
and where there are no boulders or other troubles except getting rid
of the muck; to do this it will be necessary to construct a ditch
such as the Miocene at Nome, at a large cost, but when once in operatin
will produce around two and one half or three million dollars a year
for many years to come. I have spent a goodeal of time looking into
the proposition and would be pleased to talk it over with you; I am
here at St. Louis and will be here some time; if would be possible
for you to take a run down this way sometime soon I will be glad to
explain the whole matter to you. I have maps &c. and have taken some

options on some of what I consider the most valuable properties and
can get others at reasonable prices, if no inkling gets out as to
plans &c. When you gave me your address you were with the Miama Corp.
and I addressed my letter to that address, but when I got your reply
it would appear that you are now with the American Smelting and Refining Co. Won't you please inform me if you are still with the Miami
Corp. I am addressing this to their address.

          I may come on to New York and Chicago later, but I feel that I
should get in touch with you as soon as possible, and get you to come
here if I can as I believe this to be of more importance to you andt
to me than anything else at present, and will be worthwhile the time
and expense.

                    Very Respectfully Yours,

                                   J. M. Davidson

                              Hotel Majestic,
                              St. Louis,
                              Mo.

A scanned copy of James Davidson's original letter to Norman Stines concerning mining in the Fairbanks district. In this letter, Davidson alludes to previous correspondance with Stines and mentions that he is anxious to see Stines concerning a new mining proposition. Davidson estimates that his proposition will produce between $2\frac{1}{2}$ to 3 million dollars a year for "many years to come."

return a profit there as it had in Nome. Eventually, Hammon succeeded in gaining control of 511.5 acres on Goldstream and Engineer creeks as well as an additional 456.5 acres on Cleary Creek.[8] At this time, the companies competed with one another, yet it appeared neither company could succeed alone. It was essential that Stines and Hammon reached some sort of agreement because they each faced the same problem: it was not economic for either of them to assemble a dredge on a single claim, and then have to disassemble it and move it to another claim for reassembly when the first claim was played out.

In May 1924, Stines and Hammon decided to work together, and the F.E. Company was created. First, the men combined their claims in the Fairbanks district. Next, they determined how much stock in the newly formed F.E. Company each of their own companies would receive. The Pioneer Company received 2,498 shares of F.E. stock while USSR&M received 2,497 shares and an additional $75,000 in cash. The following year, USSR&M bought out Hammon's share of the F.E. Company for $274,780. This is how F.E. became a subsidiary of USSR&M Company for $110 per share.[9]

F.E. bought out many small miners in the Fairbanks area to acquire the land they needed for dredging.[10] The company was careful to maintain a good relationship with residents of Fairbanks. John C. Boswell, a USSR&M employee, and author of the book, *History of Alaskan Operations of United States Smelting, Refining and Mining Company*, wrote that the company "enjoyed good public relations with prospective sellers and generally reached early agreements satisfactory to both parties. From our prospecting we were frequently in the position of knowing more about the value of a claim than the owner. The company offer was based on drilling results. The owner had the option of accepting the offer or leasing the claim. If the owner had an exaggerated idea of what the claim was worth, a lease was often the most satisfactory from the standpoint of both parties."[11]

Patience was required. Stines calculated that it would take about five years of preparation to acquire a water supply and a thawing system sufficient for the large dredges he wanted to bring to Fairbanks. Stines estimated that it would take an additional nineteen years to exhaust the ground that had been proven to be dredgable. By the time the first few

dredges were completed in 1928, the F.E. Company had become the largest contributor to the economy in Fairbanks and would remain so for the next thirty years.[12]

Stines estimated that it would take roughly $10 million to make gold dredging in Fairbanks a reality, with a good sum of that money being funneled directly into the community.[13]

Stines' Original Estimate of Startup Costs	
Land:	$1,716,000
Dredging Equipment:	$2,900,000
Power Plant/Lines:	$800,000
Ditch to tap Chatinika River:	$1,500,000
Fairbanks Camp:	$355,000
Camps on Creeks:	$444,000
Auxiliary Equipment:	$450,000
Secondary Ditches:	$100,000
Working Capital:	$1,500,000
Overhead:	$400,000
Total:	**$10,165,000**

Figure 5-1 is based on information from, *"The Northern Gold Fleet: Twentieth-Century Gold Dredging in Alaska,"* by Clarke C. Spence.

The *Fairbanks Daily News-Miner* was impressed by the magnitude venture and published the following article on December 24, 1925.

Waste Gold: A Story of the Second Coming of a Camp by One Who Came, Saw and Was Convinced

By: Margaret C. Deyo

It is much too early to write the history of the Fairbanks Exploration Company. Histories for the most part are written in the past tense, and most of the history of the Fairbanks Exploration Company is still in the future. But when it is finally written – as of course it will be – it will be wholly a narrative of facts viewed in the abstract by a writer whose observations have not been influenced by personal contact with its earlier developments and developers.

His or her knowledge will be garnered from the company's records, from minute books of directors' meetings and finally from financial statements showing profits against costs of operation converted into interest on investment. In short, its history will be largely a financial report and its importance will be determined by the amounts shown to be profits.

Its physical development will be therefore easily traced by anyone who has access to these records. But of its developers, the men who saw in the abandoned claims the hillsides the gold that made possible these records, there will be no word beyond mention of their terms of service, the capacities in which they served, their remunerations, and so forth. These records will contain nothing of the personality of the man whose dynamic energy and boundless optimism pointed the way to the recovery of this "waste gold," bringing prosperity and renewed life to a dying camp. Nor will these records differ in that respect from the records of any other big enterprise.

Personalities are incongruous companions of facts; but to the people of Fairbanks the personality of Norman C. Stines is so much a part of the Fairbanks Exploration Company that disassociation is impossible. To them the Fairbanks Exploration Company and Norman C. Stines are synonymous.

As before stated, any history of the Fairbanks Exploration Company at this time would be premature and presumptuous, but it is permitted to give a brief chronology of the events to date, together with a summary of what is proposed.

As early as 1920, on his return to his native America, after the Soviet government had confiscated the mines he had developed in Russia and of which he was managing director, Norman C. Stines turned his attention to Alaska and really began what is now the Fairbanks Exploration Company.

The post-war depression that had darkened all Europe had cast its shadow on American business also and particularly on metal markets. The age-old law of supply and demand had again proven its supremacy and there was over production of almost all metals except gold. And now that peace had come there must be gold to pay the indemnities.

Remembering that gold is where you find it, he turned his attention to these places. He knew that every placer field that has produced large quantities of gold by hand methods is a potential dredging or hydraulic field. If the conditions are such that dredging or hydraulicking can be done it is almost a maxim that as much gold will be recovered by these operations as has already been produced by hand.

But for dredging or hydraulicking there must be water – a plentiful and dependable supply.

Inquiries showed Nome to have the largest production of placer gold. Then came the Fairbanks district.

In 1920 Mr. Stines made a trip to Nome, studied the field and returned to New York to finance it.

In 1921 J.M. Davidson came to him and told him he believed that water could be brought into the Fairbanks district and that he intended to go in and make a preliminary study. He promised to bring any resulting business to Mr. Stines.

In 1923 Mr. Davidson came to Fairbanks, studied the water situation and took some options; in November he wired Mr. Stines suggesting a meeting in Chicago. A study of Davidson's data convinced Mr. Stines that Fairbanks should develop into a very large dredging field.

In 1907 W.P. Hammon and the American Exploration Company, the forerunner of the U.S.S.R.&M.E. Co., had examined the feasibility of building a ditch down the left limit of Chatanika to bring water into Cleary, Little Eldorado, Dome Vault and Goldstream Creeks.

Subsequently the engineers making that examination concluded it could not be done at a sufficiently low cost to warrant the undertaking. They further reported that they did not consider the frozen areas of gravel covered by great thicknesses of overburden as susceptible to dredging. That ended that proposal.

But W.P. Hammon, mindful of the knowledge gained in that examination, together with the advances made in thawing ground by the cold water method, and his experiences at Nome, again returned to this field and in 1922 sent R.H. Humphrey in to secure options on sufficient ground to warrant prospecting.

That year prospecting by drilling was done on Fish, Fairbanks, Cleary and Goldstream. This gave favorable results and was continued in 1923. Mr. Hammon did not see fit to exercise his options becoming due in the autumn of 1923, despite these results. He renewed some and dropped others.

In February, 1924, having taken over Davidson's options, Mr. Stines came to Fairbanks and took additional ones on Fish, Ester, Fairbanks, Goldstream and Cleary, and commenced drilling. His advent brought competition with Hammon company and after some negotiations the properties controlled by Hammon and those controlled by the United States Smelting and Refining Exploration Company were combined in what is now that Fairbanks Exploration Company.

Mr. Stines took charge of the new company and drilling and prospecting continued under his direction. Assisting him from the beginning was Crosby E. Keen, who had spent many years in and around Fairbanks and was familiar with all the areas under option. Later he was given full charge of all field work and made resident manager.

The results of this prospecting were such that a survey to determine the cost of bringing water to these areas was warranted.

The idea of bringing a large quantity of water by a long ditch from both the Chena and Chatanika rivers came from J.M. Davidson. The preliminary survey of the Chena proposal showed that to be impracticable. The question then became one of utilizing the Chatanika alone to serve all areas but Fish and Fairbanks.

The suggestion of tunneling from the head of Vault Creek to Fox gulch, whereby the Chatanika water would be available on Goldstream and could later be taken to Ester, came from Fred Searles, Jr.

A survey showed this to be feasible, and from that the present proposal took form.

But a water system eventually over one hundred miles long required a great deal of money to construct—therefore a really large potential supply of recoverable gold had to be available to warrant such an expenditure.

It took vision and optimism to see that much gold in those abandoned claims and engineering skill to find it. And it required courage and self-confidence to persuade a board of directors to provide the necessary funds.

Opposing Mr. Stines' views were men who, lacking his vision, considered his scheme too vast. They suggested utilizing the local waters, which Mr. Stines had measured and found insufficient and erratic.

Arguing with him against any substitution of his scheme were Mr. Davidson and Mr. Keen. Consulting engineers came and went, their findings and suggestions only served to strengthen the opinions of this trio.

Plainly it presented distinct features, not to be adjusted by precedent. WATER – that was the great need. Water under pressure.

Persistently they held to this, while prospecting results continued to prove the large expenditure justifiable. Present indications are that this persistency has won for them.

The scheme proposed means building at once 80 miles of water system; one and a quarter miles of which will be tunnel, seven miles of which will be siphons of 48 and 52 inches diameter with heads up to 505 feet, and 72 miles of which are canal proper; the erection of a 5000-kilowat power plant; the construction of six or seven dredges and the building of the necessary shops and camps to maintain the dredges and house the crews.

It includes the removal by stripping some 7,000,000 cubic yards of muck and the thawing and dredging of some 5,500,000 cubic yards of gravel each season.

It is expected that five years will be required to complete the construction program. If started in 1926 it should be completed in 1930. The first dredges would be producing in 1928.

The amount of money that will be spent to complete this work is variously estimated to be from nine to ten million dollars and no return can be expected until most of it has been spent.

To authorize such an expenditure and in the face of opposing forces speaks volumes for the vision and judgment of the directors and is indisputable evidence of their faith and reliance in the man who brought the business to their attention. That time will prove the wisdom of their faith no one in Fairbanks doubts.

From 7:30 a.m until 11:30 p.m., Norman C. Stines can be found on the job, either in the office poring over maps, drill records, and estimates, working out the general lines to be followed as well as the detail, or in the field getting acquainted with his personnel and following the work.

Ten, or even nine millions of dollars is a great deal of money. Its investment involves much responsibility, for money cannot think. It is a tremendous influence in all our lives, establishing our living standards, limiting our knowledge, even determining our health, yet this stupendous force must be directed. Wisely directed its contribution to the sum of human happiness is unlimited. Nor can its importance to Fairbanks be overestimated.

Its coming will give this camp from 20 to 40 years of prosperity. Out on the creeks now dotted with decaying log cabins and rotting tailing dumps – ugly souvenirs of the original owners who nearly all took themselves and their gold Outside – giant dredges will be in operation for many years to come, furnishing employment to all who want it and happiness to those who are willing to work for it.

Modern, electrically equipped homes will replace tumble-down log huts, a water, and sewerage system will be substituted for the well and its neighboring cesspool.

All this is according to present plans and without possible expansions and expansions may reasonably be expected to follow.

The USSR&M Company brought Joseph P. Lippincott to Fairbanks to help design what soon would be named the Davidson Ditch. Previously, Lippincott had helped engineer the 250-mile Los Angeles Aqueduct system. Because it was important that the Davidson Ditch avoid permafrost, Lippincott routed it on southward-facing slopes.[14] The Davidson Ditch was built with a .04% grade[15] so that it did not require any pumps to keep the water moving.

In 1926, several hundred men began working on the Davidson Ditch under the watchful eye of George W. Metcalfe, who had succeeded Stines, and was now in charge of all of the F.E. Company activity.[16] Much preparation was required. The first step was to remove a layer of moss and rock that ranged in depth from six inches to two feet. Removing the moss revealed a layer of muck varying from solid ice to silt. To thaw and remove the muck, crews used the same Bucyrus steam and diesel shovels that had been used to build the Panama Canal. At times, the muck would thaw too quickly, requiring the crews to shovel it by hand.[17]

After being shipped from San Francisco to Seward, the 3,000 tons of pipe for the Davidson Ditch was moved to Fairbanks on the Alaska Railroad. The pipe then was transferred to the Tanana Valley Railroad, where it was carried to Chatanika. Finally, a crew of twenty-six men with six Caterpillar tractors hauled the pipe to different destinations along the Chatanika River.[18]

In 1925, the winter snowfall was lighter than usual, causing piping to freeze to the ground. In 1927, crews had to work outdoors at temperatures of -56°F.[19]

"The term 'ditch' does not do the project justice because, although the open earthwork section comprised most of its length, the course included a 0.7-mile-long tunnel near Fox, and 6.13 miles of inverted siphons along the way," Larry Gedney wrote for the Alaska Science Forum in 1983.

The project required a lot of manpower. In 1926, the F.E. Company set up a winter camp in Chatanika capable of housing eighty men. It could feed as many as 130 men in two sittings. Next, the company built four camps for seasonal use at Gilmore, Upper Cleary Creek, Fox, and Lower Goldstream. The bunkhouses on Gilmore, Upper Cleary, and Fox were able to serve as many as 75 men at a time while the bunkhouse on Lower Goldstream served as many as 150. To feed everyone, each bunkhouse served an early and a late breakfast. If employees were unable to return to camp for lunch, they were given a thermos of coffee and packed lunch before they left for work. Bunkhouse dinners always included vegetables, two kinds of meat, and dessert. The F.E. Company maintained a ratio of one mess hall employee per every ten men boarding at their camps.[20]

In 1927, crews had to work outdoors at temperatures of -56°F.

In 1928, F.E. Company completed the Davidson Ditch and then assembled the first two dredges. Original estimates had projected that the ninety-mile Davidson Ditch would cost $1,794,294.[21] However, due to environmental factors specific to Alaska, its final cost was $3,042,745.82.[22]

Length of the Davidson Ditch	
Earth Work Section	83.27 miles
Penstock & Penstock Flumes	0.40 miles
Siphons	6.13 miles
Tunnel	0.70 miles
Total	**90.50 miles**

Figure 5-2 is based on information from, *"The Northern Gold Fleet: Twentieth-Century Gold Dredging in Alaska,"* by Clarke C. Spence. It shows the total length of the Davidson Ditch as well as the length of its parts.

Of the many people who helped build and maintain the Davidson Ditch, Leonard Seppala is probably the most famous. Seppala worked for USSR&M as a ditch superintendent in both Nome and Fairbanks.[23] Seppala later became famous helping to carry antitoxin to Nome by dog sled during a diphtheria epidemic in 1925. Not only did he carry the antitoxin over the most dangerous leg of the journey, he completed his leg in record-breaking time.[24] The serum run inspired organization of the annual Iditarod Trail Sled Dog Race from Anchorage to Nome.

Douglas K.N. Fullerton

A picture of the front and back side of a stock certificate issued by Hammon Consolidated Gold Fields on June 1, 1922 courtesty of the Fairbanks Gold Archive Collection. Hammon owned the Pioneer Mining Co. This note matured in July of 1926. In 1925, USSR&M bought out Hammon's share in F.E. for $110 per share.

Fairbanks Gold Archive Collection

A photograph of the piping network leading to Dredge No. 8. One end of the pipe was connected to the Davidson Ditch while the other could be lengthened as needed. The pipe supplied water to both the dredge and the thaw points in front of it.

Chapter Six

Building Gold Dredge No. 8

In 1927, the USSR&M Company opened bids for its first seven dredges, which were to range in size from four-foot to ten-foot dredges. They received bids from Bucyrus, Yuba Manufacturing Company, and the Bethlehem Steel Company. Bethlehem outbid the competition, but built only five dredges. The contract required that Bethlehem furnish the hulls, supply machinery, and build the superstructure. USSR&M supplied all timber, wiring, pumps, motors, and dredge buckets. The total cost for this part of the project was $1.5 million.[1] The first five dredges were named 2, 3, 5, 6 and 8, but they were not built in that order. The final three dredges were 4, 7, and 10. Unlike the first five dredges, the last three had hulls made by Yuba of pontoons that were bolted together.[2]

Even though two companies built the dredges, their purpose was the same – to extract gold buried as deep as 200 feet underground and to recover gold left behind by Fairbanks's original miners. Because many of the claims purchased by the F.E. Company already had been drift-mined, results from prospect drilling were a bit erratic. Undeterred, the F.E. Company lived by the rule that "any ground that had been good enough for drifting could be dredged profitably."[3]

Gold Dredge No. 8 was shipped in pieces from Pennsylvania to San Francisco via transcontinental railroad. On April 18, 1927, the dredge left San Francisco on the steamship *Tanana*. After its arrival in Seward, and shipment to Fairbanks by rail, an additional 2,700 feet of railroad tracks had to be laid to get it to Gilmore. No. 8 was assembled in record time, thanks to a skilled crew and good weather. On August 25, USSR&M Company took over operation of Gold Dredge No. 8.[4]

Because the dredges were powered by electricity, the F.E. Company built its own coal power plant near downtown Fairbanks. Coal was supplied by mines in Healy and brought to Fairbanks on the Alaska Railroad. Power lines were extended to the dredges. The men who maintained power lines were called Polecats. In the winter, when the dredges were shut down, the F.E. Company sold excess power to the city of Fairbanks, which had high electric demands during the dark winter months. In 1953, the F.E. Company sold its power plant to the Golden Valley Electric Association. From that point on, the F.E. purchased power from GVEA during summer months.[5]

The F.E. Company headquarters were just off of Illinois Street in Fairbanks. On one side of the street the company built a large machine shop where it repaired or built parts for the dredges. Nearby was a building that housed the "gold rooms" where all of the assaying and retorting took place. Across the street, the company built homes for its top employees. All of these historic homes still stand where they were built, though they are now privately owned.[6]

Maria Reeves

A photograph of the building that was once USSR&M headquarters taken at -30°F. Since 1953, it has been owned by GVEA. The building was restored in 2007.

Chapter Seven

Prepping the Land

Mining in a cold climate requires much more effort than it does in a warm climate. Alaskan miners had to endure harsh weather conditions that limited the season. Before Gold Dredge No. 8 could be operated, crews prepared the land in front of it by clearing a path for it to follow. This involved thawing permafrost and removing enough overburden to expose the gold-bearing land.

In order to determine what route Gold Dredge No. 8 would follow, prospecting was done with Keystone drills. The drills were spaced between 200-400 feet apart to determine the gold content of the land.[1] Because the ground was frozen, the holes were stable without support, which meant that no casings were needed. The data allowed the F.E. Company to calculate the volume of muck to be stripped as well as the volume of the gravel to be thawed. It also estimated the dollar value of the gold to be mined. At the end of each season, the company prepared a production estimate for the following season.[2] Once the route was determined, the land had to be physically prepped for dredging, which involved removing frozen overburden to a depth of 200 feet.

The percentage of a mineral found in an ore deposit determined its grade – high, medium, or low.[3]

Overburden exists in two forms: muck and silt. The difference between muck and silt is that muck is frozen while silt is not. They can be found together or separately, though muck is most commonly found in streams and valleys.[4] Muck is black and contains a large assortment of organic material, both from vegetation and animal life. Silt is most commonly found on ridge tops or upper valley walls and is light in color. It usually has very little organic material in it.[5]

Hydraulic giants were used to clear the overburden in a process called stripping. They sprayed water at rates ranging between 50 to 150 pounds per square inch.[6] As the water washed away the top layer of frozen ground, it exposed more frozen ground beneath it. The combination of water and warm weather allowed the ground to melt. Once melted, it was washed away and then the cycle was repeated. Due to Alaska's cold climate, stripping could only occur for a period of four to five months each season.

The hydraulic giants were spaced apart so that their radiuses formed a pattern of interlocking circles. It took between 36 and 48 giants to clear 30 to 40 acres of land. Usually, individual giants were operated on one or two shifts a day though they were capable of operating 24 hours a day. How often a giant was used was dependant on the material.[7]

Muck thawed very slowly. It took about two and a half hours to thaw one inch of freshly exposed muck. Unfortunately, thawed material often insulated the muck, causing it to thaw at an even slower rate. It could take as long as eight hours to thaw an average of three inches of muck. Often it would take another 16 hours to thaw through the next inch.[8]

Once the overburden had been removed, the gold-bearing land had to be thawed. Usually the land was not colder than 30°F, so thawing could be done with water at a natural temperature. Water was driven into the ground with heavy pipes attached to steel points. The points had to extend all the way to bedrock, which was often found at a depth greater than 45 feet.[9] The points were driven into the ground in a pattern of equilateral triangles. If the bedrock occurred at a depth greater than 45 feet, the points were spaced 32 feet apart from one another and thawed at a rate of 2.5 days per foot of frozen ground. If the bedrock occurred at a depth less than 45 feet, the points were spaced only 16 feet apart and thawed at a rate of 1.5 per foot of frozen ground.[10] It could take weeks for each point to reach bedrock, but once the ground was thawed it would not freeze over again.

Workers called "point drivers" drove points into bedrock with a driving clamp and a ten-pound weight.[11] The point drivers were responsible for rows of 20-30 points at a time. They continuously passed between points to take advantage of thawing that occurred between drivings.[12] With each pass, the points were thrust deeper into the ground. Once the points reached bedrock, a man called the "point doctor" checked each point and supervised the flow of water until all of the ground was thawed. Points were driven by hand up until the end of World War II, after which mechanized point drivers were used.[13]

In 1936 and 1937, the USSR&M Company experimented with an underground hydraulic system that could be useful in extracting gold from land that was too deep to dredge or too expensive to drift mine. Previous attempts by the Whitworth Mining Company and a man named Jack Howell to develop this system had failed for unknown reasons. The USSR&M Company set up a temporary camp in 1936 called, "Chatanika Flats-Shaft No.1." First, they dug a shaft that reached gravel at a depth of 52 feet. Next, two crews of two men thawed the gravel at a rate of about five feet per day. They hit bedrock at a depth of 172 feet. After that, an eight foot by eight foot section was excavated to a depth of 180 feet, and the gravel was sluiced. Gold recovery totaled 13.65 ounces and the dirt was valued at $3.32 per square foot. After the thawing process, the pay gravel and bedrock were the only materials to be treated using the hydraulic process.[14]

By 1937, the engineers who had worked on the project concluded, "The method of Underground Hydraulicking of placer gravels as used at Chatanika Flats-Shaft No. 1 was successful. The mining venture was not a success financially nor was the whole area mined that was laid out. It is believed that with what has been done and the knowledge gained, further experiment is warranted. Much improvement can be made on equipment and method of attack. The experiment proved many unknown facts; however, the short period of actual operation leaves many problems yet to be solved."[15]

UAF Archives

A photograph of a point doctor that was taken as he supervised the flow of water in one of the thaw points ahead of a dredge. Point doctors supervised water flow even after the invention of automatic point drivers.

Chapter Eight
Pleistocene Fossils

The removal of overburden in the Fairbanks district resulted in the recovery of a large variety of Pleistocene fossils. Professor Otto William Geist of the University of Alaska was responsible for recovering fossil remains as well as identifying, preserving, and cataloguing them.[1]

At times, Geist and his crew waited days for specimens to be freed by the combination of water and sun. USSR&M had a formal agreement with both UA and Childs Frick, honorary curator of the American Museum of Natural history. Most nozzlemen were careful not to damage bones as they were exposed. The only problems occurred when visitors tried to leave with fossils. Sometimes Geist had to seize the fossils from visitors who had a "finders-keepers" attitude.[2]

Hundreds of tons of pleistocene fossils were unearthed on F.E. land. Fossils included remains from bison, super-bison, lions, elks, bears, wooly mammoths, mastodons, musk oxen, horses, giant ground sloths, camels, and saber-toothed tigers.[3] Geist's most famous discovery was a baby wooly mammoth that was unearthed in 1948. The mammoth had almost been washed away before the nozzleman saw it. The nozzleman figured it was someone's sheepskin jacket, so he retrieved it, and saved it for the "bone hunters."[4]

Top Right: John Reeves and friend display a six foot long ivory mammoth tusk they unearthed with a hydraulic giant in Fox, Alaska. For video footage of the excavation, go to: http://www.youtube.com/user/mammothquest.

Bottom Right: Professor Geist and an unidentified child pose with a fully intact mammoth skull. Several complete bison skulls are visble beneath their feet.

Maria Reeves

UAF Archives

The following is an excerpt from the book *Aghvook, White Eskimo, Otto Geist and Alaskan Archaeology*, by Charles J. Keim that is quoted in John C. Boswell's *History of Alaskan Operations of United States Smelting, Refining and Mining Company*.

Each spring, prior to and after World War II, despite a winter of hard work, Otto eagerly resumed his operations "on the creeks." Quite often his and his assistants' discoveries were far from routine. One spring the hydraulic giants uncovered the skull and skeletal remains of a super-bison that evidently had washed down from a hill to Little Eldorado Creek near Chatanika. Soon the giants uncovered another, then another and yet another, until 48 such skulls and other remains were scattered in an area no more than thirty-five yards square. Otto theorized that the animals had sought food on the higher, more exposed elevations, had died, and then had been washed to the lower elevation.

Richard Osborne, an assistant working at Fairbanks Creek, discovered a foot which thousands of years earlier had been torn from the body of a young wooly mammoth and then preserved in the permafrost. The foot was complete with hoof and full growth of hair and warm undercoat of wool to protect it from the winter cold.

On another occasion, Otto retrieved the almost complete mummy of a bison from Fairbanks Creek. The specimen had wool, hide, hair, muscle, and meat fibers, some fat, a complete skull with horn cores and shells, and hoof shells. Both the mammoth leg and bison mummy were so dehydrated that they did not possess actual flesh similar to that of the mammoth that, many years earlier, had been found buried in ice in Siberia. Dogs and humans had eaten part of it.

One of Otto's most renowned finds was the partial body – head, eyeballs, trunk, and leg – of a small, wooly mammoth similar to the one found in Siberia. The hydraulic giant had badly ripped the specimen before the operator had seen it so Otto carefully sewed the specimen together, embalmed it, and then sent it air express to the Frick Laboratory.

He firmly believed that because much of the Arctic had been unglaciated, great numbers of animals traveled back and forth over the land bridge connecting Siberia and Alaska. Some Pleistocene fossils had been located in the Arctic.

Otto hoped that someday he or his assistants would discover the body of a Pleistocene-era man or at least his bones in the frozen muck. Yet, he conceded, "I believe the best possibilities for finding such human remains will be in the caves where they haven't been disturbed."

Chapter Nine

The Crew

The crew that worked at Gold Dredge No. 8 was typical of all of the dredges run by the F.E. Company. Fourteen men were required to run the dredge. They were divided into three shifts, some of which overlapped. In 1931, a man named William M. Rosewater surveyed USSR&M's operations and recommended that one more oiler and one more deckhand be added to each shift. The F.E. Company adopted Rosewater's changes.[1]

	Pre-1931	Post-1931
8 hour shifts	3 Winchmen 3 Deckhands 3 Oilers	3 Winchmen 3 Deckhands 6 Oilers
10 hour shifts	1 Dredgemaster 2 Deckhands	1 Dredgemaster 2 Deckhands
12 hour shifts	2 Deckhands	4 Deckhands
Total	14 Men	19 Men

Figure 9-1 is based on information from the book, "*The Northern Gold Fleet: Twentieth-Century Gold Dredging in Alaska,*" by Clarke C. Spence. It shows how shifts at Gold Dredge No. 8 were divided pre-1931 and post-1931.

One dredgemaster oversaw all three shifts and planned the route. The dredgemasters had extensive mining experience and often were recruited from the Lower 48 states. They were one of two employees with a key to the cleanup room. In 1934, the dredgemaster earned $380 per month without board or $330 with board included.[2]

Jobs such as that of dredgemaster or winchman required certain skills, as did any jobs requiring electricians or machinists. Even though skilled employees often were brought in from other locations, the F.E. Company recruited many employees locally.[3]

The winchman controlled all of the levers in the winchroom, located about twenty feet above the deck. The winchman monitored the devices that recorded the depth, the amount of current, and weight of the buckets. He also monitored the speed of the bucketline to make sure they hadn't hit any obstacles. The winchman had an opportunity to demonstrate his proficiency every time the dredge began a new cut. It was said that "a good winchman could time everything so that as he raised one spud near the end of the dredge's last swing, he would drop the other spud in a manner that allowed the buckets to cut along the face without having to halt the digging process."[4] Winchmen were stereotyped as being hard driving, hard drinking, and hard fisted.[5]

> Winchmen were stereotyped as being hard driving, hard drinking, and hard fisted.

The engineer did not have an engineering degree. Instead, the engineer was in charge of working with the engines. Occasionally he would act as the oiler or flume tender.[6]

The oiler was charged with maintaining the machinery.[7] He was in charge of all of the equipment as well as servicing the pumps, motors, main drive, and tumblers. Prevention was key. A good oiler could minimize breakdowns by spotting problems before they occurred.[8]

There were always two deckhands onboard. The bow deckhand kept the deck clean and broke off any bedrock that stuck to the lips of the buckets. He also would perform miscellaneous tasks as needed. The stern deckhand's job was to survey the stacker and check the rotary screen constantly.[9]

Shoremen were required to handle supplies and set shorelines.[10]

The panner was a university student that F.E. hired on. He was in charge of sampling the gravel as the dredge continued to move forward. His duties also included maintaining gold-saving tables and collecting data on ground conditions. Information that was important to the panner included the depth of bedrock, the width of the cut, the depth of muck and gravel. The panner also monitored the amount of bedrock that the dredge dug each day. He also monitored cross-sections at the height of the bank and tailing piles. The panner's reports were important to surveyors who drafted progress maps.[11] Every day, the panner entered the cleanup room to set mercury into the mercury traps. The panner was the second employee that F.E. trusted with a key to the cleanup room. F.E.'s philosophy was that the panner would not compromise future employement with the company by stealing gold. The day the panner graduated, F.E. took away the key.

Maria Reeves

Top Right: The view from inside the Winchroom. Not only was the Winchman constantly making adjustments to levers, he also kept a watchful eye on the buckets and made sure the dredge was level.

Bottom Right: The gold-saving tables inside the cleanup room. Mercury traps and riffles line the sluices.

Maria Reeves

Chapter Ten

Safety

Although most injuries were minor, working on any dredge had the potential to be a very dangerous activity. The most frequent injuries were bone fractures, hernias, or fragments lodging in the eye. Seven fatalities occurred during the F.E. reign. These fatalities ranged from drownings to workers being pinned inside a Caterpillar tractor after frozen gravel fell. One man died when the hydraulic giant he was operating spun out of control and crushed his chest. In 1932, eight accidents were reported on Gold Dredge No. 8.[1]

First-Aid classes were held at many of USSR&M camps where safety teams were selected to respond to accidents. On August 22, 1930, a safety contest was held in Fox. Five safety teams competed in Farthest North Safety Championship before a crowd of 150 people. The competition resulted in a tie between Chatanika and Goldstream camps. The outcome was decided by a coin toss, which Chatanika won.[2] These contests were created to build interest in safety and to inspire responsibility. Boswell wrote that most accidents were caused by "careless acts of which all of us are guilty. Let's 'WATCH OUR STEP.'"[3]

F.E. Accident Reports: 1928-1930				
Year	Serious Accidents	Minor Accidents	Total Accidents	Man Shifts
1928	31	209	240	168,344
1929	34	134	168	128,097
1930	15	110	125	113,777

Figure 10-1 is adapted from the book, "*History of Alaska Operations of United States Smelting, Refining And Mining Company*," by John C. Boswell. The number of reported accidents dropped dramatically in 1930.

Chapter Eleven

Gold Dredge No. 8:
The Tour

Gold Fever's Tale

I am the spirit you mortals call fever.
Gold Fever's my name, if you please.
I live in the mountains, the valleys and hills,
I live in the sands of the seas.

Africa, Asia, Alaska, Peru,
I flourish and thrive in each place.
Some of the men I lured got rich,
and some of them were a disgrace.

Alaska was only a study I did
to see how far men could exceed
their limit of strength and wills to survive
and quench their feverish greed.

Fairly successful so far I would say.
I've been bested by mortals up here.
They conquered the death like cold and the land
and vanquished the dark and the fear.

The fight was a mean one, but I stacked the deck,
threw the worst that I had in their path.
The muck and frost and mosquitoes did naught
but strengthen their wills and their wrath.

Look at the land beneath where you stand
and gaze at the hills all around
There's nothing more there than dirt spiked with gold
and smothered with ice-frozen ground.

My dear Mother Nature and I had a pact –
we'd make Alaska our home.
She'd pepper her bounty and beauty about,
and I'd salt the creeks as we'd roam.

Some while back we had us a tiff.
Mother was displeased for some reason.
Had something to do with the treasures I'd hid
and something to do with the season.

We argued a little and swore, just a bit,
then fought as two bears in a cage.
That crafty Old Lady then settled the score
and brought on a global ice age.

All of the treasures I'd buried with care
and all of my plans were postponed;
she covered them over with glaciers of ice
and waited till I had atoned.

We talk of it now on occasion and laugh,
the works that we planned and designed.

The treasures she placed on top of my own
are cherished like no other kind.

Her artworks are beauty, her mountains and lakes;
the creatures live wild and free.
Including the mortals I lured to this land,
the ones who came searching for me.

My story is a simple one –
You've heard that Alaska gets cold,
that life's greatest gamblers came to get rich,
the miners in search of the gold.

Boomtowns sprang up like weeds in a field,
near valleys with gold-laden soil,
and the last boom we had was a new one up North,
exploring and drilling for oil.

Alaska is vast and its rivers run deep,
I planted their souls with my seed.
They found that wealth was a measure of luck,
fanning the flames of their greed.

The valleys are quiet and peaceful,
the tundra is moss-covered ice,
and some of those leaving Alaska one time
won't be able to go twice.

The mountains refresh a downtrodden spirit
or heal a soul in despair,
imprison your thoughts when you're all alone,
and capture a glimpse or a stare.
The thrill of adventure brought most of them North.

The glory was here to pursue,
and when it got old or stale for them,
another began all anew.

The Klondike in Dawson, the Chilkoot Pass –
they fought to survive or they died,
paid God with prayers on their way up the pass,
then picked his pockets on the other side.

Families and sweethearts were left far behind
crying a half world away;
"Alaska, you thief," they'd whimper at night,
"Come home soon," they would pray.

The trails were teachers; their lessons were cruel.
Alaska taught men to survive.
And most of them did, by the skin of their teeth,
and most of them learned how to thrive.

Stories are told of these old-timers' wills,
adventurers brazen and bold,
their plight in uncharted wilderness,
their endless quest for the gold.

Each one is true; I'll tell you a few –
authentic as well, I do pledge
from aboard the machine that lured them here –
The Number 8 Bucketline Dredge.
                    – John Reeves

Gold Dredge No. 8 was the third dredge to be manufactured by Bethlehem Steel's shipbuilding division. Its design incorporated a ship-like steel hull 99 feet long and 50 feet wide. The hull was ten feet, six inches deep. Fully loaded, her seven-foot, nine-inch draft displaced 1,065 tons, including the ballast of steel machinery and on-board equipment. A coal-fired boiler produced steam, which powered the internal generators, air compressors, and pumps. This also provided heat under the conveyor belts and other vital locations to prevent permafrost-chilled gravel and sand from freezing to the moving parts.[1]

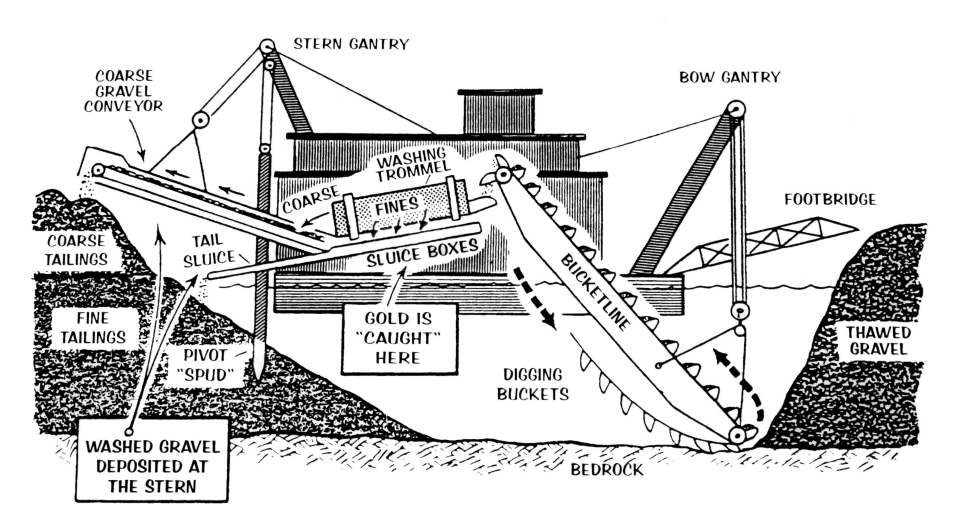

Figure 11-1 is a diagram of Gold Dredge No. 8 published courtesy of Holland America Westours. The Bucketline brought dirt into the dredge where it passed through the trommel. Anything larger than the holes in the trommel passed out of the dredge via the stacker, where it became part of the tailing pile labled "coarse tailings." Anything smaller than the holes in the trommel passed into the sluice boxes where the gold was "caught." Anything that was not caught by the sluice boxes passed through the tail sluice and was deposited at the base of the tailing labeled "fine tailings." Even today there is little vegetation on No. 8's tailing pile due to the reversal of topsoil.

All of the controls were located in the winch room and controlled by the winchman. The winchman constantly adjusted levers and speed controls to keep the dredge digging efficiently. His main purpose was to ensure that the buckets always scraped bedrock. If the buckets missed bedrock by even a few inches, it meant that gold was being left behind. The winchman also had to ensure that the dredge remained level so that the sluice boxes, also known as the gold-saving tables, would not lose gold.

Just outside and in front of the winch room, was the bucketline. It was similar to a bicycle chain because it constantly revolved around a steel digging ladder. Each manganese-steel bucket would dig into the

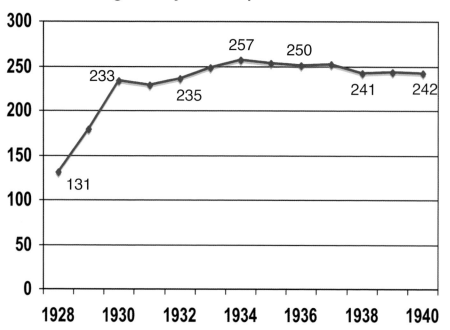

Average Days of Operation: 1928-1940

Figure 11-2 is based on information from the article "*Operations of United States Smelting, Refining & Mining Company,*" which was printed in an issue of "*Mining World*" in 1942. It is a chart with the average number of days that all of the dredges operated between 1928 and 1940. The averages between 1928 and 1930 are lower than later years because some of the dredges were under construction during part of the season.

> It was said that the F.E. Co. was so thorough with its operations "that a workman that fell into the dredge hopper came out of the stacker without a scratch – but minus his four gold teeth."

face of the pond, and pick up a load of gravel. Next, the full bucket would travel up the bucketline and dump out all of the dirt into the hopper, a black box known as the dump box. Once the bucket was empty, it would continue back down the underside of the bucketline and repeat the process. Gold Dredge No. 8 had a distinctively steep bow gantry, towering forty-three feet high, to support the bucketline.

Sixty-four buckets were attached to the bucketline, each with a capacity to hold six cubic feet of dirt. Each empty bucket weighed 1,583 pounds. The bucketline was capable of digging to a depth of 35 feet below the surface of the pond.

At top speed, the bucketline emptied 22 buckets per minute. It was important that the buckets moved constantly because every minute that the dredge was idle, the F.E. Company was losing up to 22 unrecoverable buckets. Gold Dredge No. 8 had the capacity to dig through approximately 6,000 cubic yards of gravel in a 24-hour period. The company operated at 90% time-efficiency during its first six seasons, an impressive record even by today's standards. It was said that the F.E. Co. was so thorough with its operations "that a workman that fell into the dredge hopper came out of the stacker without a scratch – but minus his four gold teeth."[2]

An average dredging season in Alaska lasted 253 days, beginning in March and ending in November. In 1930, thanks to good weather, one of the F.E. dredges set a season record by operating 269 days. The company earned $3,912,600 dredging in 1930, almost $1,000,000 more than it earned in 1929. Even though weather played an important role, most of the increase was attributed to expanding operations.[3]

In advance of the dredge, draglines were used to scrape the gravel down to bedrock. All of the dirt the draglines moved eventually was run through a sluice box. Currently, there is a massive twelve-cubic-yard dragline on display in front of Gold Dredge No. 8 gift shop.

Behind the winch room were giant gears powered by the motor on the bottom deck. These gears enabled the bucketline to move. Gravel was dumped into the hopper located behind these gears.

Maria Reeves

A photograph of the dragline currently on display at Gold Dredge No. 8. This particular dragline is capable of holding twelve cubic yards of gravel. Draglines were used to scrape the gravel down to bedrock.

Maria Reeves

A photograph of one of the gears inside of Gold Dredge No. 8. This gear is twelve feet in diameter and is made out of chrome-nickel steel. It is unaffected by temperature fluctuations and resistent to oxidation.

UAF Archives

Chatanika Gold Dredge No. 3 in action. A 10-foot dredge, No. 3 was one of the largest dredges owned by the F.E. Co. While smaller in size, Gold Dredge No. 8 had 64 buckets on its bucketline and consumed 9,000 gallons of water every minute. The bucketline on No. 8 operated at a maximum speed of 22 buckets per minute.

After the dirt was dumped into the hopper, it was carried to the trommel, a cylinder that had a diameter of over six feet and length of just over 36 feet. The trommel slopes towards the rear of the dredge and has holes in it ranging in size from $^3/_8$ to $1^5/_8$ of an inch. As the gravel was carried through the trommel, the trommel rotated slowly. A spray bar inside the trommel spewed 9,000 gallons of water per minute, breaking apart the gravel.

Gold Dredge No. 8 consumed 9,000 gallons of water per minute, 24 hours per day, seven days a week for an average season of 253 days. Add to this water consumed by the seven other dredges and it was no wonder that the Davidson Ditch was necessary. Local water sources and streams were not reliable enough.

Maria Reeves

The trommel inside Gold Dredge No. 8. has a diameter of six feet and is just over 36 feet in length. During operation, the trommel rotated around a spray bar that was used to break the gravel apart.

Everything larger than the holes in the trommel, including large gold nuggets, would continue up a conveyor belt that ran inside the stacker.

The trommel was unable to differentiate between gold and waste material. Anything smaller than the holes in the trommel would fall through into the gold-saving table below. Everything larger than the holes in the trommel, including large gold nuggets, would continue up a conveyor belt that ran inside the stacker. The conveyor belt was 32 inches wide and 262 feet long. Eventually, all of the waste material was spewed out of the stacker, creating tailing piles behind the dredge as it moved along. These tailings were useful. As the dredge moved forward, the tailings would fill in the pond behind it, causing the pond to move with the dredge.

Even many years later, little vegetation is found on the tailing piles. The reason plants struggle to grow on tailings is that dredging the land reversed the topsoil. All of the good soil was washed out of the sluice boxes at the base of the dredge only to be covered by the large rocks that fell out of the stacker.

Fairbanks Gold Archive Collection

Occasionally clay lodged itself inside the trommel. At one point, Dredge No. 2 sank because the crew decided to dislodge such a piece of clay with dynamite. Fortunately, the F.E. Co. was able to raise and repair the dredge.

Maria Reeves

The stacker from inside Gold Dredge No. 8. The tailing pile is visible to the right. Although No. 8 stopped operating in 1959, there is little vegetation on the tailing piles to this day. The conveyor belt inside the stacker was 32 inches wide. Tailing piles could be stacked at a maximum height of 27 feet above the water line.

There are two cleanup rooms located beneath the trommel, one on each side. The cleanup rooms were also known as the gold-saving tables. Each room was a series of eight sluice boxes that slanted toward the outer wall of the dredge. Three components helped the gold-saving tables catch gold – mercury traps, wooden riffles, and coco matting.

Mercury traps were thick, rubber mats with holes gouged in them. These mats lined the top third of each sluice box. Every day, the water supply would be turned off so that the panner could enter the gold-saving tables and fill the mercury traps with liquid mercury. Once Gold Dredge No. 8 was operating, a mixture of mud, silt, sand, small rocks, black sand, and gold would fall through the trommel screen. As the mixture of gravel washed over the traps, the mercury would bond with the gold to form a hard, white substance called an amalgam. Mercury would form an amalgam only with gold. Everything else would be washed away.

The riffles, a series of bars designed to catch gold not caught in the mercury traps, lined the bottom two-thirds of each of the sluices on the upper level. They completely lined the sluices on the lower level. Gold would sink and become trapped between the slats of wood while the lighter waste would wash out of the back of the dredge.

The extremely coarse coco mats were made from the fiber of coconuts. Their sole purpose was to catch black sand, which was almost as heavy as gold and often had tiny bits of gold attached to it. The coco mats were collected and stored until winter, when the dredges were shut down and labor could be spared to collect the gold from these mats.

Maria Reeves

Maria Reeves

Left: Mercury traps lined the top third of each sluice box. The mercury formed a bond called an amalgam with any of the gold that it touched. During the cleanup process, the amalgam would be separated from the gold in a process called retorting.

Right: Riffles lined the lower two-thirds of each sluice box. The riffles were designed to catch gold as it passed over them. Because gold is so dense, it became caught in the riffles while everything else washed over it and out the bottom of the dredge.

PICTURE OF A ROCKER BOX

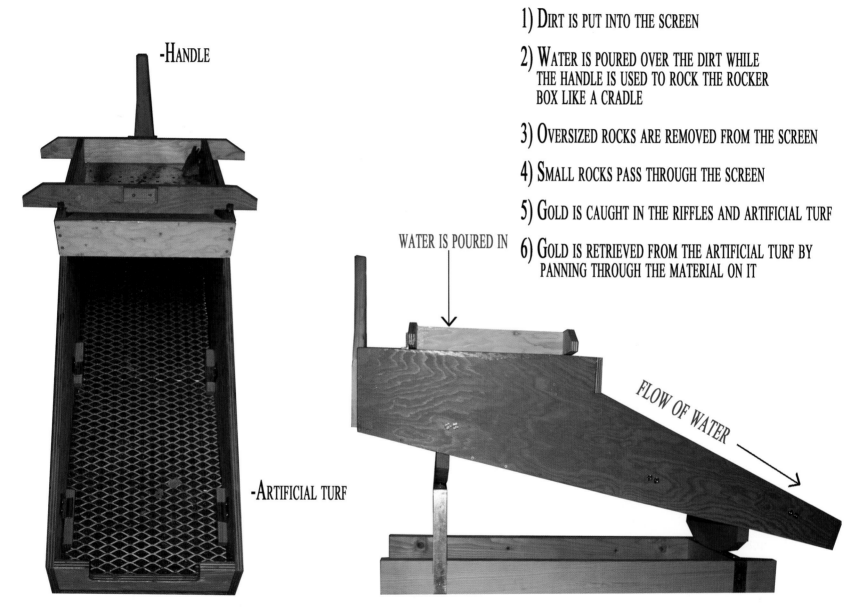

1) DIRT IS PUT INTO THE SCREEN

2) WATER IS POURED OVER THE DIRT WHILE THE HANDLE IS USED TO ROCK THE ROCKER BOX LIKE A CRADLE

3) OVERSIZED ROCKS ARE REMOVED FROM THE SCREEN

4) SMALL ROCKS PASS THROUGH THE SCREEN

5) GOLD IS CAUGHT IN THE RIFFLES AND ARTIFICIAL TURF

6) GOLD IS RETRIEVED FROM THE ARTIFICIAL TURF BY PANNING THROUGH THE MATERIAL ON IT

-HANDLE

-ARTIFICIAL TURF

WATER IS POURED IN

FLOW OF WATER

Maria Reeves

This is the front and side view of a Rocker Box that has been built to specification by John Reeves from original USSR&M blue prints. It is five feet in length and drops 1.6" per foot. Most modern Rocker Boxes drop between 1"-2" per foot. Even though this model was used in the early 1900s, its design is still accurate by today's standards. Rocker boxes are similar to sluice boxes, like the gold-saving tables in No.8, but require less water.

The F.E. Company did not have to worry too much about theft because the only employees with a key to the cleanup room were the panner and dredgemaster. In 1934, Franklin D. Roosevelt's New Deal put the United States onto the gold standard[4], which inadvertently deterred theft. The New Deal was meant to make the federal government the only legal purchaser of gold. It also allowed the federal government to set the price at $35 per ounce. Thus, if an employee hid gold in his boots or pockets, he was unable to sell it to anyone but the federal government. There was still occasional theft, but the motive was usually limited to obtaining a souvenir. If a person was caught stealing, not only would they be fired, but they also would have a hard time finding work anywhere else in Fairbanks.

In 1934, a brick of gold weighing 700 troy ounces was worth $24,500 at $35 per ounce. In the 1990s, that same brick was worth around $245,000 at $350 an ounce. In 2008, that brick was worth $700,000 at $1,000 per ounce.

The first floor of Gold Dredge No. 8 housed a 250-horsepower motor that powered the belts that moved the gears that moved the bucketline. The dredge also had a pump with its own separate motor to suck water out of the pond and pump it up to the trommel. There was a huge stock of tools and a forge on the first deck. If a piece of equipment broke down, it was possible to bring in a replacement part from the F.E. warehouses in Fairbanks. If a replacement was unavailable, the necessary part could be forged inside the dredge. No. 8 also had separate electric motors for its jitney winches, hoist winches, jibs, cranes, and air-compressors. Most of the machinery and superstructure was enclosed and sheltered from the elements, which had the potential to be cold even during the summer.

One more important feature of the bottom deck of Gold Dredge No. 8 was the catch-all at the bottom of the hopper. After the buckets dumped their load into the hopper, they were sprayed with water to make sure they were empty when they picked up their next load of dirt. Everything that washed off of the buckets would fall into the catch-all, where the company stored it, until the material could be sluiced.

> In 1934, a brick of gold weighing 700 troy ounces was worth $24,500 at $35 per ounce. In the 1990s, that same brick was worth around $245,000 at $350 an ounce. In 2008, that brick was worth $700,000 at $1000 per ounce.

It is important to note that every single bit of dirt No. 8 picked up eventually was sluiced. Thanks to gold-saving innovations such as the coco matting and catch-all, the F.E. Company was able to excavate 8.5% more gold than prospect drilling had indicated was in the ground before dredging began.

In order for No. 8 to operate an entire season, as much as four or five feet of ice would have to be removed from the pond in the spring. Ice-cutting was done using steam cutters, which drew power from the dredge's boiler. One man would operate two to four cutters, cutting 70 square feet of ice per hour. The ice was cut into 3,500 to 5,000 pound cakes, which were hoisted out of the pond from the stern gantry with a jitney line. As much as 15,000 tons of ice would be removed from a pond in a single season.[5]

Management constantly pressured employees to maintain yardage – forward movement – even late in a season when ice began to build up on the pond. Some dredge men, to extend the season, blasted the ice with dynamite and used the dredge as a battering ram against the ice. The battering was hard on the dredge. In 1931, the F.E. Company reinforced the hull plates on dredge Nos. 2, 5, 6 and 8. The hull plates on No. 3 were overhauled in 1932.[6]

Fairbanks Gold Archive Collection

A crew of men cut the ice with steam cutters to prepare it for removal from Dredge No. 5's pond.

Fairbanks Gold Archive Collection

The ice was cut into 3,500 to 5,000 pound cakes. Spring ice removal added two months to the season.

Fairbanks Gold Archive Collection

A jitney line attached to the stern gantry was used to hoist the ice out of the pond at Dredge No. 8 in March, 1948.

Fairbanks Gold Archive Collection

Thousands of tons of ice were removed from each pond in a single season.

Chapter Twelve
Cleanup

Every two weeks, Gold Dredge No. 8 was shut down for cleanup, a time for the regular crew to make repairs while a special crew was brought in to remove the gold.

Whenever No. 8 was shut down during the season, the dredge pond would be surveyed and plotted onto a map showing the dredge's progress. Then, the actual gold recovery would be compared to the "prospect value" for both that cleanup period and the season to date. The average was calculated by dividing the recovery by the prospect value.[1]

Figures varied widely between cleanups, but as the dredged area increased, the result edged closer to 100%. From 1928 through 1964, dredges in the Fairbanks district operated at an average 108.5% of prospect value.[2]

Often, the regular crew would find themselves repairing broken or cracked bucket lips. These lips originally were riveted into place, but later it was discovered that bucket lips could be replaced more quickly if heavy-duty bolt fasteners held them on instead of rivets. It took an average of three minutes to replace the buckets on No. 8 during the relipping process.[3]

The cleanup crew removed the mercury traps and riffles. It also collected the coco mats. Next, the crew scraped the amalgamated gold into locked boxes for delivery to Fairbanks for retorting. An average cleanup for Gold Dredge No. 8 yielded about 4,000 troy ounces.

The amalgamated material then was scooped into a retort furnace in which the gold was separated from the mercury and other impurities. It was heated until the mercury turned into vapor. Then, the vapor was cooled until the mercury returned to a liquid state. The mercury was puri-fied and re-used. Once the mercury was removed from the furnace, the remaining pure liquid gold was poured into brick molds and sent to the U.S. Mint in San Francisco. The Post Office charged a special gold rate of only $18 per brick of gold.[4]

Uncredited

This is a photograph of gold that was mined by a family in Ester, near Fairbanks, that has been working the same property for three generations. They have requested that their names and the name of their creek remain unidentified.

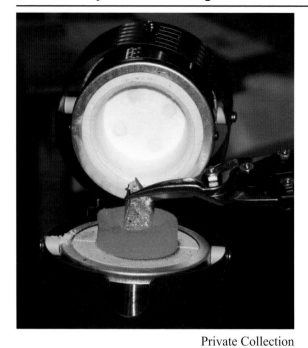

Private Collection

Gold being loaded into a melting furnace.

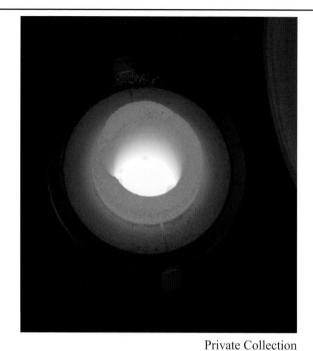

Private Collection

Gold at its melting point of 1947.52 °F.

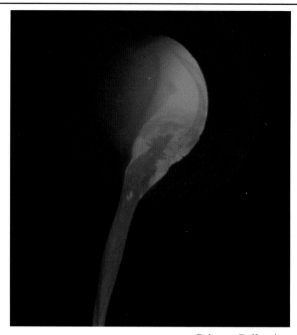

Private Collection

Liquid gold being poured into a mold.

Private Collection

Gold immediatly after being poured into a mold.

Private Collection

The bar of gold as it cools in the snow.

Private Collection

The 24.99 troy ounce bar of gold after it cooled.

Chapter Thirteen

WWII Shuts Down Gold Dredge No. 8

World War II had an effect on gold mining, not only in Alaska but also throughout the country. In 1942, the army prohibited travel to Alaska unless for defense or military work. By July 1942, the F.E. Company had only six dredges remaining in operation.[1] At least 138 F.E. employees served in the military, 23 of whom died in the war.[2]

At this time, Under Secretary of War Robert Patterson and Under Secretary of the Navy James Forrestal worked to convince the War Production Board that gold mining must be completely halted if the United States was to achieve economic mobilization for the war effort. Their argument was that gold mining not only tied up critical machinery, but also tied up supplies and diverted skilled laborers. General Brehon Somervell told the board, "Failure to stop gold production immediately would be inexcusable."[3]

The board made a ruling on October 8, 1942, known as Order L-208, prohibiting the mining of gold unless the gold was a by-product of a strategic mineral.[4] It ordered that all gold mining work be halted within 60 days, with exception to maintenance. L-208 also prohibited development of new gold mines.[5] A week later, USSR&M Company sent orders to Fairbanks to shut everything down immediately.[6]

In an annual report dated December 31, 1942 the consequences of L-208 are described along with USSR&M's plans to resume mining whenever possible. "At the Alaska gold properties, favorable development continued but the operations were shut down by government order in the fall of 1942. Maintenance crews are keeping the dredges and plants in repair and in readiness for resumption of operation when that becomes possible. Some of the buildings and equipment have been leased to the government and revenues will help to defray some of the shutdown expenses."[7]

The end of WWII did not allow dredging operations to return to full swing immediately. It was a challenge for the F.E. Company to find skilled workers and to compete with wages that the government paid on its construction projects. On top of that, the price of gold had not changed a penny from $35 per ounce after the federal government froze the price in 1934, though operating costs had increased.[8]

The F.E. Company struggled to stay in operation through the 1950s by reducing the number of operating dredges. Gold Dredge No. 8 continued to operate until 1959, when she was shut down for economic reasons.[9] Dredges No. 2 and No. 3 were the final two dredges to stop operating. They were shut down in 1964.

And, prostitute like, her hey-day was slight
With a few feverish years of success
Then the charm of her youth was faded—forsooth,
And she sat in desertion's distress!
Now, broken and worn, the object of scorn,
She bows to calamity sold,
A derelict wan, bleak, battered and gone,
Unknown to her lovers of old!
-Harold Salisbury

An aerial view of Gold Dredge No. 8, long before she was closed for L-208. No. 8 was nicknamed the "Queen of the Fleet," because she was the only dredge that spent her entire mining career operating in the Goldstream Valley.

Chapter Fourteen
The Legacy of Gold Dredge No. 8

It was the end of an era when the F.E. Company shut down its dredges for good. Yet, during their more than 30-year run, they had given Fairbanks a second chance at economic survival. Fortunately, when the dredging stopped, Fairbanks did not fade from the map. To this day, Fairbanks continues to grow, thanks mostly to a large military presence, gold production, and North Slope oil production.

During WWII, the War Department realized how important the Interior of Alaska was to national security. The Alaska-Canada Highway (ALCAN) was built in 1942-43 to improve communication and supply routes between Alaska and the Lower 48 states. The highway was completed in 18 months, allowing the military to expand both Ladd Field (later to become Fort Wainwright when the War Department was reorganized as the Department of Defense) as well as Eielson Air Force Base.

Fairbanks boomed through the 1970s when it became the hub of construction for one of the world's most amazing engineering projects of all time, the trans-Alaska pipeline spanning 800 miles from the North Slope oil fields to the ice-free tanker port of Valdez.

Meanwhile, gold mining made another comeback in the 1970s when the federal government abandoned the gold standard and lifted restrictions against private ownership of gold. Without a fixed price, gold prices rose as did the number of active gold mines. By 1982, more than 500 placer mines operated in Alaska.[1]

Gold Extraction from Alaska: 1982-1984			
Year	Gold in Ounces	Value at the Time	Value in 2008
1982	174,900	$70 Million	$174.9 Million
1983	169,000	$67.6 Million	$169 Million
1984	175,000	$63 Million	$175 Million

Figure 15-1 is from the magazine, *"The Role of Placer Mining in the Alaska Economy."* It charts the value of all gold taken out of Alaska between 1982-1984 at both its original value and its 2008 value.

Chapter Fifteen

Restoring Gold Dredge No. 8

In 1982, John Reeves realized the dredge's potential. Together with his wife, Ramona Reeves, the couple purchased No. 8 and restored it.

One of the first things Reeves did after he purchased No. 8 was relocate the bunkhouse from its original location at Goldstream Camp to its current location overlooking the dredge. When asked what he paid for the bunkhouse, Reeves reluctantly answered that he was pretty sure it was somewhere between "fifty cents and a buck and a half." His $1 investment was contingent on his moving the two-story, 12-bedroom building to its new location about a half mile away. The grueling move lasted a few hours, though preparation took months.

The Fairbanks Daily News-Miner printed the following article on May 5, 1984.

Fox Dredge Wins Historic Status

Another national historic site has been added to Fairbanks' inventory with the formal notice that Gold Dredge No. 8 has been accepted, and the owners say they plan to expand it as a tourist attraction.

John and Ramona Reeves purchased the five-deck, 250-foot long dredge and site off 9 Mile Steese Highway in late 1982, and recently were notified that it has been designated for the National Register of Historic Places.

Last year the Reeves opened the dredge to tours. Admission is $5. Visitors each receive a small piece of gold overlay from the site and have an opportunity to pan for gold, keeping any nuggets they find. Guides emphasize the history of mining, the gold fields, and Fairbanks.

New, beginning in June, will be an evening offering of "Dredge O'Drama," a locally written play featuring three characters. It will be included as part of after-dinner tours on Friday, Saturday, and Sunday evenings. Visitors can bring their own refreshments or purchase beverages and desserts, said Reeves.

Gold Dredge No. 8 was built in 1928. Its 12-bedroom bunkhouse, which once accommodated 50 miners, was built in 1926. It operated at its present location until 1959, when it was shut down. The original owner was the U.S. Smelting and Refining, and its subsidiary, Fairbanks Exploration Co., said Reeves.

A group of Fairbanks businessmen later purchased the dredge and in the early 1970s offered tours, but that enterprise ended about 1973.

Reeves believes this is the only gold mining dredge in Alaska that is privately owned and available for public viewing.

From the deck "you can see modern mining operations and the (oil) pipeline," Reeves says, and the dredge is surrounded by the hills of Goldstream Valley.

Several local tour companies last year began including a visit to the dredge as part of their offerings to groups.

Reeves said he hopes in future years to add a small hotel in the bunkhouse, a restaurant, bar, and gift shop. A 4,000-square foot deck is being built this summer.

Chapter Sixteen

Gold Dredge No. 8 Today

Though the city of Fairbanks has changed immensely over the years, Goldstream Valley has changed little since Gold Dredge No. 8 ceased working in 1959. The surrounding hills still are covered by alder and birch trees. For countless visitors, this historic dredge provides a superb experience and honors Alaska's mining heritage. No. 8 is still the best and most easily accessible landmark of its kind in Alaska. It retains its "gold feeling" and gives visitors the opportunity to experience some of the magic that helped spur Fairbanks from a primitive gold camp into the modern city it is today. Visitors are welcome to explore the property and pan for gold in the panning area.

Maria Reeves

The bunkhouse gift shop and part of the panning area. Reeves restored and moved the bunkhouse gift shop to its current location in 1992.

Maria Reeves

Two theaters and a bathhouse on site at Gold Dredge No.8. The theaters regularly show the movie, "*Alaskan Gold: Placer Mining in Interior Alaska.*"

Maria Reeves

Gold Dredge No. 8 today. Various tools, hydraulic giants and gears line the path in front of the dredge. A giant gold pan and painted nugget present a unique "photo op" in front of the dredge.

Today, the dredge has much to offer visitors. It has several theaters, warehouses, and displays easy to reach and full of information.

The theaters show a narrated video called, "Alaskan Gold: Placer Mining in Interior Alaska," produced by KUAC-TV at the University of Alaska Fairbanks. The video is a remake of a silent film made in 1949.

Several warehouses are filled with remnants and artifacts, including records, tools, and supplies abandoned when the F.E. Company shut down the dredge. The dredgemaster's house has a display dedicated to the dredging process. It shows how the overburden was removed and how power was supplied to the dredge. Another display shows how Gold Dredge No. 8 was shipped to Alaska.

The bathhouse has been left almost completely untouched. More than 75 men shared the tiny coal-powered bathhouse where they bathed and washed clothes.

The bunkhouse museum has four rooms and an entryway. Inside is a reconstruction of the miners' sleeping quarters as well as historical information about gold rushes in the Fairbanks area. Also displayed is a detailed list of duties of the dredge crew.

The Hallway of Bones is a large display of Pleistocene remains. The American Museum of Natural History granted thousands of dollars to USSR&M Company to have the bones shipped to their laboratory. USSR&M was the only company with a formal commitment to collect fossils they unearthed for scientific study.[1]

Maria Reeves

The inside of the bathhouse on display at Gold Dredge No. 8. At one time, 75 miners shared this coal-powered bathouse.

Maria Reeves

The Dredgemaster's Mess Hall. Every summer day, a hearty Miner's Stew Lunch is available to visitors.

Chapter Seventeen

Alaskan Nuggets of Note

Since 1880, more than 33 million troy ounces of gold have been mined in Alaska. Due to its density, that amount of gold would fill only about one-third of a boxcar on the Alaska Railroad.[1]

The creeks near Fairbanks combined to yield the richest paystreaks in the state of Alaska at eight million ounces. The Nome area provides the second richest paystreaks in Alaska at 7.5 million ounces.[2]

Miners still routinely recover nuggets ranging from one to fifteen troy ounces.[3] Often miners recover large nuggets using a simple gold pan and shovel, though it is also possible to recover nuggets with a metal detector. Fairbanks remains the richest gold district in the state of Alaska, and it is still relatively easy to find gold in the area today.

Top Ten Nuggets Recovered in Alaska

Troy Ounces	Name	Area	Year	Found By
294.1	"The Alaskan Centennial Nugget"	Ruby	1998	Barry Clay
182	"The Anvil Nugget"	Nome	1903	Pioneer Mining Co.
146	N.A.	Wiseman	N.A.	N.A.
137	N.A	Wiseman	N.A.	N.A.
127	N.A.	McGrath	N.A.	N.A.
122	"The Ganes Nugget"	McGrath	1985	Lloyd Magnuson
107	N.A.	Nome	N.A.	N.A.
97	N.A.	Nome	N.A.	N.A.
95	N.A.	Nome	N.A.	N.A.
92	"The Heart of Gold"	Kantishna	1985	Mic Martinek

Figure 18-1 has been compiled based on information from the book, "*Alaska Gold: A Prospector's Guide,*" by Ron Wendt as well as the website, www.akmining.com. It is a list of the top ten largest nuggets recovered in Alaska as of January 2009.

Chapter Eighteen

How to Gold Pan

Gold panning is a relatively easy skill to learn. An experienced gold panner can work through a pan of dirt in a few minutes while a novice may take much longer. No matter how long the process, the goal is to efficiently find and recover as much gold as possible.

In Alaska, there are many places where a visitor can pan for gold. However, it is important to pay attention to your surroundings and respect private property during your quest. In general, Alaskans warn off trespassers with signs or gates.

The first step is to submerge the pan and shake it for a few seconds. As you shake the pan, the larger rocks will collect at the surface because they are less dense than the other materials in the pan.

Next, gently fill the pan with water and either set the pan down or hold it still in one hand. Now, you can wash the large rocks over the pan before you remove them. The reason for this step is that gold can be hidden in any of the mud or dirt clinging to the large rocks. If you forget to wash the rocks over your pan before you toss them aside, there is a good chance that you will lose gold. Note that every time you want to remove debris from your pan, whether it is a rock, branch, or root, it is a good idea to wash it over the pan first.

Once the larger debris has been removed, submerge your pan and shake it again. Eventually, you will have a layer of small stones that has collected at the surface. Instead of painstakingly picking the stones out one by one, submerge the pan at about a 45-degree angle and gently dip the pan in and out of water. Allow the water to wash the top layer of stones away. You should not worry about losing gold during this process, though you should keep an eye out for larger nuggets during this phase. Once the top layer has washed away, repeat the entire process beginning with submerging and shaking the pan. Repeat this step until only a fine layer of sand remains in your pan.

The final step is the most fun. The fine layer of sand remaining at the base of your pan will consist of other dense materials such as hematite, black sand, and garnets. At this point, let a small amount of water into your pan and shake the pan at an angle so that all of the material ends up at the top of the pan. You can hold the pan with one or both hands. Holding the pan with both hands is more stable and easier for the novice. No matter what your style, slowly rock the pan so that the water moves around it in a slow, circular motion. All of the lighter material will wash to the bottom of the pan while any gold flecks or nuggets will remain at the top of the pan. The gold will remain still as the water passes slowly over it.

Ramona Reeves

A young Maria Reeves pans for gold. Reeves grew up at Gold Dredge No. 8 and caught "gold fever" around the same time she learned to walk.

About the Author

I loved growing up at Gold Dredge No. 8, which is why I had so much fun writing this book. My four siblings and I never were short of things to do at the dredge. My favorite activities included playing in the panning area, searching for mammoth bones, and exploring. We spent a good deal of time swimming in various ponds and playing in an enormous sand pit that since has been filled in.

When I was in elementary school, my class would take field trips to my house in the spring. I always had a great time taking my friends on tour and trying to teach them how to pan for gold. At the end of the day, the school buses would come to pick everyone up and I stayed behind, waving everybody off. I was always a bit confused why we didn't make field trips to anybody else's house.

I joined tours of the dredge every chance I got. The ultimate tour guide was my dad, and I especially loved hearing about the gold-saving tables. I have an active imagination and still enjoy thinking about different forms of thievery employed by some on the cleanup crew. Whether they were true or not, it was fun to hear about people shoveling amalgamated gold with hollow shovels or hiding gold dust by running their gold-covered fingers through their 1950s greased hair.

My parents are hard workers, and I caught on to that at a young age. They are also smart and used their talents to make many improvements to the dredge and surrounding buildings. Somehow, they convinced me that cleaning the public restrooms for $2 a day was a fair deal. While it was not as fun as "three kids lemonade," the stand operated by my little sister, brother and I, it was steady. It eventually paid for a class trip to Paris and also got me used to getting out of bed early, which later helped me in my swimming career.

Douglas Fullerton

Maria Reeves smiles for the camera during an ice climb in Fox, Alaska.

I could go on all day with stories that took place while my family lived at Gold Dredge No. 8, like F/A-18s shattering every single window in the bunkhouse. Or sledding face first into a dredge bucket – they're hard to see when they're covered in snow! Unfortunately, I will have to save these stories for another book.

I hope you have enjoyed reading about Gold Dredge No. 8 as much as I have enjoyed writing about her.

Happy prospecting,

Maria Reeves

Glossary

Aggredation
Aggredation occurs when the amount of sediment in a river is larger than what the river is able to transport. The sediment is deposited by the river and results in an increase in elevation.

Alloy
An alloy is a mixture of metals that are usually melted together in order to improve their qualities. Gold alloys are used in jewelry and are stronger than gold alone.

Alluvial Deposit
An alluvial deposit is created when rivers deposit sediment.

Amalgam
An amalgam is formed when a substance reacts with mercury. Mercury was used to "catch" gold by forming an amalgam with the gold in the mercury traps. The mercury was separated from the gold in a retort furnace and then re-used.

Avoirdupois
The Avoirdupois system is a non-metric system used in the United States to measure mass. An avoirdupois ounce weighs 437.5 grains. An avoirdupois pound weighs sixteen ounces.

Bedrock
Bedrock is the name of the rock that forms the earth's crust. It can be igneous, metamorphic or sedimentary. The ground above bedrock forms the foundation for plant and animal life.

Bench Deposit
A bench deposit is formed as a river system cuts through land. As continued erosion occurs, the bench remains elevated above a former or current riverbed.

Black Sand
Black sand is a type of fine sand that is found in placer deposits. Like gold, black sand is dense. Heavy concentrates of black sand are often found with gold.

Bucketline
The bucketline on a dredge constantly rotates like a bicycle chain. The buckets are full of material as they travel up the bucketline. At the apex of the bucketline, the buckets dump their material into the hopper before continuing down the bucketline where they scoop up more material. Dredges are classified by how many cubic feet of dirt their buckets held.

Bunkhouse
A bunkhouse is the type of building that was used to house F.E. Company employees. The bunkhouse at Dredge No. 8 has 12 bedrooms upstairs. Four guys shared each room, but only two men were ever in a room at the same time. The shifts were designed so that two men slept while two men worked at a time.

Coarse Gold
Gold that weighs more than one milligram is defined as Coarse gold.

Creeping
Creeping is a slow process that occurs as rock and soil advance down a low-grade slope.

Dragline
A dragline is an enormous bucket that is suspended from a boom with wire or rope. The dragline is used in a process called stripping to move overburden. One of the draglines on display at Gold Dredge No. 8 has the capacity to hold twelve cubic yards of material.

Dredgemaster
The dredgemaster is the dredging equivalent to the captain of a ship. The dredgemaster oversaw the shifts and planned the route.

Gold Fever
The stampeders who came north for the gold rush were afflicted with gold fever.

Gold Purity
Gold purity in the United States is measured in Karats. 24 Karat gold is pure, which means that it has 999.99 parts gold per thousand.

Gold Standard
A gold standard is a standard of currency that is based on gold. In 1934, Franklin D. Roosevelt put the United States on the gold standard. Richard Nixon took the United States off of the gold standard in 1971.

Hematite
Hematite is the mineral form of Iron oxide. Because of its density, it is often found alongside gold. Hematite has a metallic luster.

Karat

A Karat is a system of measurement used to define the purity of gold and gold alloys.

Keystone Drill

Keystone drills were used by USSR&M to prospect the land and determine what route the dredges would take.

Lode Gold

Lode gold is found inside hard rock formations and has to be extracted by crushing the rock surrounding it.

Melting Point

The temperature at which a solid changes state from a solid to a liquid is its melting point. Gold has a melting point of 1947.52 °F.

Mercury

Mercury is one of six elements that are liquid at room temperature and pressure. Mercury was used in the gold-saving tables because it forms an amalgam with gold.

Muck

Muck is frozen. It contains a large assortment of organic material, both from vegetation and animal life. Muck is usually found in streams and valleys.

Nozzleman

The men who operated the hydraulic giants in front of the dredges were called nozzlemen.

Order L-208

On October 8, 1942 Order L-208 was passed. L-208 prohibited the mining of gold unless it was recovered as a by-product of a strategic mineral. L-208 ordered that all mining work be stopped within 60 days of its issue, with exception to maintenance. It also prohibited the development of new gold mines.

Overburden

Overburden is the word used in mining to describe all of the material that lies above bedrock. At times up to 250 feet of overburden had to be removed to expose bedrock.

Permafrost

Permafrost is land that is permanently frozen.

Placer Gold

Placer deposits result from natural weathering. They can be deposited by rivers or carried by glaciers.

Pleistocene Fossil

Pleistocene fossils date from 1.8 million to 10,000 years ago. The Pleistocene epoch ended with the retreat of the last continental glacier.

Point Doctor

The man who supervised the flow of water through thaw points was called the point doctor.

Point Driver

Point drivers were the men responsible for hand-driving thaw points into the ground with a ten-pound weight. Point drivers were responsible for rows of 20-30 points at a time. They continuously passed between points to take advantage of thawing that occurred between drivings

Polecat

The men who maintained power lines were called polecats.

Quartz

Quartz is the most abundant mineral on the earth's continental crust. Often gold is found within quartz veins.

Retorting

Retorting is the process in which mercury is separated from gold.

Seward's Folly

Secretary of State, William Seward, engineered the Alaska Purchase. At the time, it was quite unpopular, so it was nicknamed Seward's Folly.

Sheet Wash

Sheet wash is a slow form of erosion in which a thin film of water transports soil over the ground.

Silt

Silt is commonly found on ridge tops or upper valley walls. It is light in color and usually has very little organic material in it.

Specific Gravity

Specific gravity is a measure of density. Substances with a specific gravity greater than one sink in water while those with a specific gravity lighter than one float in water.

Spud

The two spuds in the rear of the dredge were used as anchor points to prevent the dredge from moving. Without spuds, the natural movement of the buckets would have pushed the dredges backwards. The spuds on Gold Dredge No. 8 weighed many of tons and jsut below fifty feet in length.

Tertiary Period

The Tertiary Period is made up of five epochs that span between 65 million and 1.8 million years ago. The Tertiary marked the start of the Cenozoic Era. At the beginning of the Tertiary, mammals became the dominant vertebrates.

Troy Ounce

A troy ounce is a unit of measurement that is only used to describe the mass of precious metals. A troy ounce is made of 480 grains and there are twelve troy ounces in one troy pound.

Bibliography

Chapter Two

[1] John Reeves. Personal interview. November, 2008.

[2] John Reeves. Personal interview. November, 2008.

[3] L.J Campbell, "Alaska's Mineral Industry Today," Alaska Geographic 22.3 (1995): 32.

Chapter Three

[1] The Fairbanks Placer Gold Deposits (With Map Folio), J.M. Metcalfe and Ralph Tuck, July 1940. 5.

[2] The Fairbanks Placer Gold Deposits (With Map Folio), J.M. Metcalfe and Ralph Tuck, July 1940. 5.

[3] The Fairbanks Placer Gold Deposits (With Map Folio), J.M. Metcalfe and Ralph Tuck, July 1940. 5.

[4] The Fairbanks Placer Gold Deposits (With Map Folio), J.M. Metcalfe and Ralph Tuck, July 1940. 17.

[5] The Fairbanks Placer Gold Deposits (With Map Folio), J.M. Metcalfe and Ralph Tuck, July 1940. 21-23.

[6] The Fairbanks Placer Gold Deposits (With Map Folio), J.M. Metcalfe and Ralph Tuck, July 1940. 145-147.

[7] The Fairbanks Placer Gold Deposits (With Map Folio), J.M. Metcalfe and Ralph Tuck, July 1940. 77.

[8] The Fairbanks Placer Gold Deposits (With Map Folio), J.M. Metcalfe and Ralph Tuck, July 1940. 73.

[9] The Fairbanks Placer Gold Deposits (With Map Folio), J.M. Metcalfe and Ralph Tuck, July 1940. 149.

[10] The Fairbanks Placer Gold Deposits (With Map Folio), J.M. Metcalfe and Ralph Tuck, July 1940. 159-160.

[11] L.J Campbell, "Alaska's Mineral Industry Today," Alaska Geographic 22.3 (1995): 43.

[12] The Fairbanks Placer Gold Deposits (With Map Folio), J.M. Metcalfe and Ralph Tuck, July 1940. 177.

[13] John Reeves. Personal interview. November, 2008.

[14] Mike Webb, master jeweler and partner at Gold Rush Fine Jewelry. Personal interview. November, 2008.

Chapter Four

[1] L.J Campbell, "Alaska's Mineral Industry Today," Alaska Geographic 22.3 (1995): 38.

[2] Terrence Cole, Crooked Past (Fairbanks: University of Alaska Press, 1991) 13-16.

[3] Terrence Cole, Crooked Past (Fairbanks: University of Alaska Press, 1991) 21-26.

[4] Terrence Cole, Crooked Past (Fairbanks: University of Alaska Press, 1991) 15-25.

[5] Dermot Cole, Fairbanks: A Gold Rush Town that Beat the Odds (Canada: Kent Sturgis, 1999) 15-16.

[6] Dermot Cole, Fairbanks: A Gold Rush Town that Beat the Odds (Canada: Kent Sturgis, 1999) 16-18.

[7] Dermot Cole, Fairbanks: A Gold Rush Town that Beat the Odds (Canada: Kent Sturgis, 1999) 17.

[8] Terrence Cole, Crooked Past (Fairbanks: University of Alaska Press, 1991) 40.

[9] Terrence Cole, Crooked Past (Fairbanks: University of Alaska Press, 1991) 42-44.

[10] Terrence Cole, Crooked Past (Fairbanks: University of Alaska Press, 1991) 52-60.

[11] Terrence Cole, Crooked Past (Fairbanks: University of Alaska Press, 1991) 67.

[12] Terrence Cole, Crooked Past (Fairbanks: University of Alaska Press, 1991) 14-15.

[13] R.N. DeArmond, qtd. in Terrence Cole, Crooked Past (Fairbanks: University of Alaska Press, 1991) 39.

[14] Dermot Cole, Fairbanks: A Gold Rush Town that Beat the Odds (Canada: Kent Sturgis, 1999) 26.

[15] L.J Campbell, "Alaska's Mineral Industry Today," Alaska Geographic 22.3 (1995): 39.

[16] The American Society of Mechanical Engineers. "Gold Dredge Number 8, National Historic Mechanical Engineering Landmark," ASME May 1986: 2.

[17] Dermot Cole, Fairbanks: A Gold Rush Town that Beat the Odds (Canada: Kent Sturgis, 1999) 63.

[18] Dermot Cole, Fairbanks: A Gold Rush Town that Beat the Odds (Canada: Kent Sturgis, 1999) 63-68.

[19] The American Society of Mechanical Engineers. "Gold Dredge Number 8, National Historic Mechanical Engineering Landmark," ASME May 1986: 2.

Chapter Five

[1] The Alaska Mining Hall of Fame Foundation, "James M. Davidson," The Paystreak 3.1 (2001): 5-6.

[2] Clarke C. Spence, The Northern Gold Fleet: Twentieth-Century Gold Dredging in Alaska (Urbana and Chicago, University of Illinois Press, 1996) 75.

[3] The Alaska Mining Hall of Fame Foundation, "James M. Davidson," The Paystreak 3.1 (2001): 5-6.

[4] The Alaska Mining Hall of Fame Foundation, "James M. Davidson," The Paystreak 3.1 (2001): 5-6.

[5] The Alaska Mining Hall of Fame Foundation, "Norman C. Stines," The Paystreak 3.1 (2001): 4.

[6] The Alaska Mining Hall of Fame Foundation, "Norman C. Stines," The Paystreak 3.1 (2001): 4.

[7] Clarke C. Spence, The Northern Gold Fleet: Twentieth-Century Gold Dredging in Alaska (Urbana and Chicago, University of Illinois Press, 1996) 75.

[8] Clarke C. Spence, The Northern Gold Fleet: Twentieth-Century Gold Dredging in Alaska (Urbana and Chicago, University of Illinois Press, 1996) 76.

[9] Clarke C. Spence, The Northern Gold Fleet: Twentieth-Century Gold Dredging in Alaska (Urbana and Chicago, University of Illinois Press, 1996) 76.

[10] John Boswell, History of Alaskan Operations of United States Smelting, Refining and Mining Company (Fairbanks: Mineral Industries Research laboratory, 1979) vii.

[11] John Boswell, <u>History of Alaskan Operations of United States Smelting, Refining and Mining Company</u> (Fairbanks: Mineral Industries Research laboratory, 1979) 14.

[12] John Boswell, <u>History of Alaskan Operations of United States Smelting, Refining and Mining Company</u> (Fairbanks: Mineral Industries Research laboratory, 1979) vii.

[13] Clarke C. Spence, <u>The Northern Gold Fleet: Twentieth-Century Gold Dredging in Alaska</u> (Urbana and Chicago, University of Illinois Press, 1996) 76.

[14] Clarke C. Spence, <u>The Northern Gold Fleet: Twentieth-Century Gold Dredging in Alaska</u> (Urbana and Chicago, University of Illinois Press, 1996) 78.

[15] John Boswell, <u>History of Alaskan Operations of United States Smelting, Refining and Mining Company</u> (Fairbanks: Mineral Industries Research laboratory, 1979) 11.

[16] Clarke C. Spence, <u>The Northern Gold Fleet: Twentieth-Century Gold Dredging in Alaska</u> (Urbana and Chicago, University of Illinois Press, 1996) 79.

[17] Clarke C. Spence, <u>The Northern Gold Fleet: Twentieth-Century Gold Dredging in Alaska</u> (Urbana and Chicago, University of Illinois Press, 1996) 79.

[18] Clarke C. Spence, <u>The Northern Gold Fleet: Twentieth-Century Gold Dredging in Alaska</u> (Urbana and Chicago, University of Illinois Press, 1996) 79.

[19] Clarke C. Spence, <u>The Northern Gold Fleet: Twentieth-Century Gold Dredging in Alaska</u> (Urbana and Chicago, University of Illinois Press, 1996) 80.

[20] Clarke C. Spence, <u>The Northern Gold Fleet: Twentieth-Century Gold Dredging in Alaska</u> (Urbana and Chicago, University of Illinois Press, 1996) 186.

[21] Clarke C. Spence, <u>The Northern Gold Fleet: Twentieth-Century Gold Dredging in Alaska</u> (Urbana and Chicago, University of Illinois Press, 1996) 78.

[22] Clarke C. Spence, <u>The Northern Gold Fleet: Twentieth-Century Gold Dredging in Alaska</u> (Urbana and Chicago, University of Illinois Press, 1996) 80.

[23] John Boswell, <u>History of Alaskan Operations of United States Smelting, Refining and Mining Company</u> (Fairbanks: Mineral Industries Research laboratory, 1979) 53.

[24] John Boswell, <u>History of Alaskan Operations of United States Smelting, Refining and Mining Company</u> (Fairbanks: Mineral Industries Research laboratory, 1979) 59.

Chapter Six

[1] Clarke C. Spence, <u>The Northern Gold Fleet: Twentieth-Century Gold Dredging in Alaska</u> (Urbana and Chicago, University of Illinois Press, 1996) 82.

[2] John Boswell, <u>History of Alaskan Operations of United States Smelting, Refining and Mining Company</u> (Fairbanks: Mineral Industries Research laboratory, 1979) 25.

[3] John Boswell, <u>History of Alaskan Operations of United States Smelting, Refining and Mining Company</u> (Fairbanks: Mineral Industries Research laboratory, 1979) 14.

[4] Clarke C. Spence, <u>The Northern Gold Fleet: Twentieth-Century Gold Dredging in Alaska</u> (Urbana and Chicago, University of Illinois Press, 1996) 84-85.

[5] Clarke C. Spence, <u>The Northern Gold Fleet: Twentieth-Century Gold Dredging in Alaska</u> (Urbana and Chicago, University of Illinois Press, 1996) 163-164.

[6] John Reeves. Personal Interview. November, 2008.

Chapter Seven

[1] John Reeves. Personal interview. November, 2008.

[2] John Boswell, <u>History of Alaskan Operations of United States Smelting, Refining and Mining Company</u> (Fairbanks: Mineral Industries Research laboratory, 1979) 14.

[3] L.J Campbell, "Alaska's Mineral Industry Today," <u>Alaska Geographic</u> 22.3 (1995): 34.

[4] John Boswell, <u>History of Alaskan Operations of United States Smelting, Refining and Mining Company</u> (Fairbanks: Mineral Industries Research laboratory, 1979) 15.

[5] The Fairbanks Placer Gold Deposits (With Map Folio), J.M. Metcalfe and Ralph Tuck, July 1940. 17.

[6] John Boswell, <u>History of Alaskan Operations of United States Smelting, Refining and Mining Company</u> (Fairbanks: Mineral Industries Research laboratory, 1979) 16.

[7] Clarke C. Spence, <u>The Northern Gold Fleet: Twentieth-Century Gold Dredging in Alaska</u> (Urbana and Chicago, University of Illinois Press, 1996) 85.

[8] John Boswell, <u>History of Alaskan Operations of United States Smelting, Refining and Mining Company</u> (Fairbanks: Mineral Industries Research laboratory, 1979) 16.

[9] Clarke C. Spence, <u>The Northern Gold Fleet: Twentieth-Century Gold Dredging in Alaska</u> (Urbana and Chicago, University of Illinois Press, 1996) 85.

[10] Clarke C. Spence, <u>The Northern Gold Fleet: Twentieth-Century Gold Dredging in Alaska</u> (Urbana and Chicago, University of Illinois Press, 1996) 85.

[11] Clarke C. Spence, <u>The Northern Gold Fleet: Twentieth-Century Gold Dredging in Alaska</u> (Urbana and Chicago, University of Illinois Press, 1996) 85.

[12] John Boswell, <u>History of Alaskan Operations of United States Smelting, Refining and Mining Company</u> (Fairbanks: Mineral Industries Research laboratory, 1979) 21.

[13] Clarke C. Spence, <u>The Northern Gold Fleet: Twentieth-Century Gold Dredging in Alaska</u> (Urbana and Chicago, University of Illinois Press, 1996) 85.

[14] John Boswell, <u>History of Alaskan Operations of United States Smelting, Refining and Mining Company</u> (Fairbanks: Mineral Industries Research laboratory, 1979) 55-56.

[15] John Boswell, <u>History of Alaskan Operations of United States Smelting, Refining and Mining Company</u> (Fairbanks: Mineral Industries Research laboratory, 1979) 55-56.

Chapter Eight

[1] John Boswell, <u>History of Alaskan Operations of United States Smelting, Refining and Mining Company</u> (Fairbanks: Mineral Industries Research laboratory, 1979) 17.

[2] John Boswell, <u>History of Alaskan Operations of United States Smelting, Refining and Mining Company</u> (Fairbanks: Mineral Industries Research laboratory, 1979) 18-19.

[3] John Boswell, <u>History of Alaskan Operations of United States Smelting, Refining and Mining Company</u> (Fairbanks: Mineral Industries Research laboratory, 1979) 18.

[4] John Boswell, <u>History of Alaskan Operations of United States Smelting, Refining and Mining Company</u> (Fairbanks: Mineral Industries Research laboratory, 1979) 20.

Chapter Nine

[1] Clarke C. Spence, <u>The Northern Gold Fleet: Twentieth-Century Gold Dredging in Alaska</u> (Urbana and Chicago, University of Illinois Press, 1996) 166.

[2] Clarke C. Spence, <u>The Northern Gold Fleet: Twentieth-Century Gold Dredging in Alaska</u> (Urbana and Chicago, University of Illinois Press, 1996) 172.

[3] Clarke C. Spence, <u>The Northern Gold Fleet: Twentieth-Century Gold Dredging in Alaska</u> (Urbana and Chicago, University of Illinois Press, 1996) 171.

[4] Clarke C. Spence, <u>The Northern Gold Fleet: Twentieth-Century Gold Dredging in Alaska</u> (Urbana and Chicago, University of Illinois Press, 1996) 169.

[5] Clarke C. Spence, <u>The Northern Gold Fleet: Twentieth-Century Gold Dredging in Alaska</u> (Urbana and Chicago, University of Illinois Press, 1996) 169.

[6] Clarke C. Spence, <u>The Northern Gold Fleet: Twentieth-Century Gold Dredging in Alaska</u> (Urbana and Chicago, University of Illinois Press, 1996) 169.

[7] Clarke C. Spence, <u>The Northern Gold Fleet: Twentieth-Century Gold Dredging in Alaska</u> (Urbana and Chicago, University of Illinois Press, 1996) 169.

[8] Halibuk, Patrick and David Neufeld. <u>Make It Pay! : Canadian Gold Dredge #4: Klondike, Yukon, Canada</u> (Canada: Kegley Books, 1994) 41.

[9] Clarke C. Spence, The Northern Gold Fleet: Twentieth-Century Gold Dredging in Alaska (Urbana and Chicago, University of Illinois Press, 1996) 169.

[10] Clarke C. Spence, The Northern Gold Fleet: Twentieth-Century Gold Dredging in Alaska (Urbana and Chicago, University of Illinois Press, 1996) 169.

[11] Clarke C. Spence, The Northern Gold Fleet: Twentieth-Century Gold Dredging in Alaska (Urbana and Chicago, University of Illinois Press, 1996) 169-170.

Chapter Ten

[1] Clarke C. Spence, The Northern Gold Fleet: Twentieth-Century Gold Dredging in Alaska (Urbana and Chicago, University of Illinois Press, 1996) 184.

[2] John Boswell, History of Alaskan Operations of United States Smelting, Refining and Mining Company (Fairbanks: Mineral Industries Research laboratory, 1979) 52.

[3] John Boswell, History of Alaskan Operations of United States Smelting, Refining and Mining Company (Fairbanks: Mineral Industries Research laboratory, 1979) 52.

Chapter Eleven

[1] John Reeves. Personal interview. November, 2008.

[2] Clarke C. Spence, The Northern Gold Fleet: Twentieth-Century Gold Dredging in Alaska (Urbana and Chicago, University of Illinois Press, 1996) 96.

[3] Clarke C. Spence, The Northern Gold Fleet: Twentieth-Century Gold Dredging in Alaska (Urbana and Chicago, University of Illinois Press, 1996) 88.

[4] Clarke C. Spence, The Northern Gold Fleet: Twentieth-Century Gold Dredging in Alaska (Urbana and Chicago, University of Illinois Press, 1996) 71.

[5] John Boswell, History of Alaskan Operations of United States Smelting, Refining and Mining Company (Fairbanks: Mineral Industries Research laboratory, 1979) 23.

[6] Clarke C. Spence, The Northern Gold Fleet: Twentieth-Century Gold Dredging in Alaska (Urbana and Chicago, University of Illinois Press, 1996) 89.

Chapter Twelve

[1] John Boswell, History of Alaskan Operations of United States Smelting, Refining and Mining Company (Fairbanks: Mineral Industries Research laboratory, 1979) 15.

[2] John Boswell, History of Alaskan Operations of United States Smelting, Refining and Mining Company (Fairbanks: Mineral Industries Research laboratory, 1979) 15.

[3] Clarke C. Spence, The Northern Gold Fleet: Twentieth-Century Gold Dredging in Alaska (Urbana and Chicago, University of Illinois Press, 1996) 96.

[4] Clarke C. Spence, The Northern Gold Fleet: Twentieth-Century Gold Dredging in Alaska (Urbana and Chicago, University of Illinois Press, 1996) 217.

Chapter Thirteen

[1] Clarke C. Spence, The Northern Gold Fleet: Twentieth-Century Gold Dredging in Alaska (Urbana and Chicago, University of Illinois Press, 1996) 114-115.

[2] Clarke C. Spence, The Northern Gold Fleet: Twentieth-Century Gold Dredging in Alaska (Urbana and Chicago, University of Illinois Press, 1996) 114.

[3] Clarke C. Spence, The Northern Gold Fleet: Twentieth-Century Gold Dredging in Alaska (Urbana and Chicago, University of Illinois Press, 1996) 115.

[4] John Boswell, History of Alaskan Operations of United States Smelting, Refining and Mining Company (Fairbanks: Mineral Industries Research laboratory, 1979) 58.

[5] Clarke C. Spence, The Northern Gold Fleet: Twentieth-Century Gold Dredging in Alaska (Urbana and Chicago, University of Illinois Press, 1996) 115.

[6] John Boswell, History of Alaskan Operations of United States Smelting, Refining and Mining Company (Fairbanks: Mineral Industries Research laboratory, 1979) 58.

[7] John Boswell, History of Alaskan Operations of United States Smelting, Refining and Mining Company (Fairbanks: Mineral Industries Research laboratory, 1979) 58.

[8] Clarke C. Spence, The Northern Gold Fleet: Twentieth-Century Gold Dredging in Alaska (Urbana and Chicago, University of Illinois Press, 1996) 125.

[9] Clarke C. Spence, The Northern Gold Fleet: Twentieth-Century Gold Dredging in Alaska (Urbana and Chicago, University of Illinois Press, 1996) 134.

Chapter Fourteen

[1] John Reeves. Personal interview. November, 2008.

Chapter Sixteen

[1] Clarke C. Spence, The Northern Gold Fleet: Twentieth-Century Gold Dredging in Alaska (Urbana and Chicago, University of Illinois Press, 1996) 98.

Chapter Seventeen

[1] L.J Campbell, "Alaska's Mineral Industry Today," Alaska Geographic 22.3 (1995): 38.

[2] Ron Wendt, Alaska Gold: A Prospector's Guide (Wasilla, Goldstream Publications, 1988) 36.

[4] Ron Wendt, Alaska Gold: A Prospector's Guide (Wasilla, Goldstream Publications, 1988) 36.

Index